The Kinesin Superfamily Handbook

The Kinesin Superfamily Handbook

Transporter, Creator, Destroyer

Edited by
Claire T. Friel

CRC Press
Taylor & Francis Group
Boca Raton London New York

CRC Press is an imprint of the
Taylor & Francis Group, an **informa** business

First edition published 2020
by CRC Press
6000 Broken Sound Parkway NW, Suite 300, Boca Raton, FL 33487-2742

and by CRC Press
2 Park Square, Milton Park, Abingdon, Oxon, OX14 4RN

© 2020 Taylor & Francis Group, LLC

CRC Press is an imprint of Taylor & Francis Group, LLC

Library of Congress Cataloging-in-Publication Data

Names: Friel, Claire (Claire T.), editor.
Title: The kinesin superfamily handbook : transporter, creator, destroyer / edited by Claire Friel.
Description: First edition. | Boca Raton : CRC Press, 2020. | Includes bibliographical references and index.
Identifiers: LCCN 2019059121 | ISBN 9781138589568 (hardback) | ISBN 9780429491559 (ebook)
Subjects: LCSH: Kinesin.
Classification: LCC QP552.K46 K563 2020 | DDC 572/.6--dc23
LC record available at https://lccn.loc.gov/2019059121

ISBN: 978-1-138-58956-8 (hbk)
ISBN: 978-0-429-49155-9 (ebk)

Typeset in Times
by Deanta Global Publishing Services, Chennai, India

Visit the eResources: www.crcpress.com/9781138589568

Dedication

To Cara and Catherine, my little creators and destroyers!

Contents

Acknowledgements

Firstly, I would like to thank all the authors who worked with me on this book; without their efforts this project would never have come to fruition. Secondly, I would like to thank the team at the publishers, Taylor & Francis; without the initial friendly and encouraging contact from Francesca McGowan the project would never have been started, and without the continued chivvying along and moral support from Kirsten Barr and Rebecca Davies the project would never have been finished. Finally, I would like to thank my husband, my children and the members of my lab for their patience, especially in the final few weeks, when I was not as available as I should have been.

In a work of this type there is never enough space for every piece of information that could be included or for every piece of work that could be referenced. Therefore, I and my fellow authors would like to apologise to all the scientists who have increased our knowledge of the kinesin superfamily but whose work could not be included in this book.

For online material visit the eResources: www.routledge.com/ 9781138589568.

Claire Friel.

Claire T. Friel

Editor bio

Claire T. Friel earned a BSc in Biochemistry from the University of Glasgow, UK, and carried out her PhD work on protein folding kinetics in the laboratory of Sheena Radford at the University of Leeds, UK. In 2006, she joined the group of Jonathon Howard at the Max Planck Institute of Molecular Cell Biology and Genetics in Dresden, Germany. There, she solved the ATP turnover cycle of the microtubule depolymerising Kinesin-13, MCAK, and developed an interest in the kinesin superfamily of molecular motors. Since 2011, Claire has held the position of Assistant Professor at the University of Nottingham, UK. The research goals of the Friel lab are to understand the relationship between the kinesin motor domain sequence and the many functional properties of the kinesin superfamily and to understand the molecular mechanisms of proteins that regulate microtubule dynamics.

List of Contributors

Hanan M. Alghamdi
School of Life Sciences
Medical School
Queen's Medical Centre
University of Nottingham
Nottingham, UK

Hannah R. Belsham
School of Life Sciences
Medical School
Queen's Medical Centre
University of Nottingham
Nottingham, UK

Marcus Braun
Institute of Biotechnology of the Czech
    Academy of Sciences, BIOCEV
Prague West, Czech Republic

Stefan Diez
B CUBE – Center for Molecular
    Bioengineering
Technische Universität Dresden
Dresden, Germany

Hauke Drechsler
B CUBE – Center for Molecular
    Bioengineering
Technische Universität Dresden
Germany

Claire T. Friel
School of Life Sciences
Medical School
Queen's Medical Centre
University of Nottingham
Nottingham, UK

Larisa Gheber
Department of Chemistry and Ilse Katz
    Institute for Nanoscale Science and
    Technology
Ben-Gurion University of the Negev
Beer Sheva, Israel

Alina Goldstein-Levitin
Department of Chemistry and Ilse Katz
    Institute for Nanoscale Science and
    Technology
Ben-Gurion University of the Negev
Beer Sheva, Israel

Jaspreet Singh Grewal
Centre for Mechanochemical Cell
    Biology and Division of Biomedical
    Sciences
Warwick Medical School
University of Warwick
Coventry, UK

David D. Hackney
Department of Biological Sciences
Carnegie Mellon University
Pittsburgh, PA, USA

William O. Hancock
Department of Biomedical Engineering
Pennsylvania State University
University Park, PA

Zdenek Lansky
Institute of Biotechnology of the Czech
    Academy of Sciences, BIOCEV
Prague West, Czech Republic

Tianyang Liu
Institute of Structural and Molecular
    Biology
Birkbeck College
London, UK

Andrew D. McAinsh
Centre for Mechanochemical Cell
    Biology and Division of Biomedical
    Sciences
Warwick Medical School
University of Warwick
Coventry, UK

Carolyn A. Moores
Institute of Structural and Molecular
   Biology
Birkbeck College
London, UK

Alejandro Peña
Institute of Structural and Molecular
   Biology
Birkbeck College
London, UK

Mary Popov
Department of Chemistry and Ilse Katz
   Institute for Nanoscale Science and
   Technology
Ben-Gurion University of the Negev
Beer Sheva, Israel

Fiona Shilliday
Institute of Structural and Molecular
   Biology
Birkbeck College
London, UK

Nida Siddiqui
Centre for Mechanochemical Cell
   Biology & Division of Biomedical
   Sciences
Warwick Medical School
University of Warwick
Coventry, UK

Anne Straube
Centre for Mechanochemical Cell
   Biology & Division of Biomedical
   Sciences
Warwick Medical School
University of Warwick
Coventry, UK

Alison E. Twelvetrees
Sheffield Institute for Translational
   Neuroscience
University of Sheffield
Sheffield, UK

Julie P. Welburn
Wellcome Trust Centre for Cell Biology
University of Edinburgh
Michael Swann Building, The King's
   Buildings
Mayfield Road, Edinburgh, UK

1 Introduction to the Kinesin Superfamily

Hannah R. Belsham and Claire T. Friel

CONTENTS

The kinesins are a superfamily of proteins that interact with the microtubule cytoskeleton. Kinesins use the turnover of ATP to regulate their interaction with microtubules. The first kinesin was discovered in 1985 as a soluble protein that supported ATP-dependent movement of purified microtubules (Vale, Reese, and Sheetz 1985). This protein was found to be distinct from the previously identified actin-associated motor protein, myosin, and from the microtubule-associated motor protein, dynein, and was given the name kinesin from the Greek 'kinein', meaning 'to move' (Vale, Reese, and Sheetz 1985). The principal role of kinesin was initially considered to be transport of organelles (Vale, Reese, and Sheetz 1985, Vale et al. 1985a, b), although a role in mitosis was also suggested (Scholey et al. 1985). This original kinesin was shown to be a complex consisting of two kinesin heavy chains (KHCs) and two light chains (KLCs) (Bloom et al. 1988). The nucleotide- and microtubule-binding activity was soon shown to be confined to an ~45kD N-terminal region of the KHC containing the so-called 'motor domain' (Scholey et al. 1989, Kuznetsov et al. 1989). Kinesin was found in a wide variety of organisms and cell types, including avian and mammalian neuronal tissue (Vale, Reese, and Sheetz 1985, Brady 1985, Kuznetsov and Gelfand 1986); squid neural tissue (Vale, Reese, and Sheetz 1985); sea urchin eggs (Scholey et al. 1985); *Xenopus* eggs (Neighbors, Williams, and McIntosh 1988); and *Drosophila melanogaster* (Saxton et al. 1988), although it remained unclear whether kinesin was part of a multigene family. However, after the first sequence of a kinesin, the *D. melanogaster* KHC, became publicly available (Yang, Laymon, and Goldstein 1989), a large number of kinesin-like proteins were discovered by screening for sequence homology to the characteristic kinesin motor

1

domain. Kinesins are now known to be ubiquitous among eukaryotes (Wickstead and Gull 2006, Wickstead, Gull, and Richards 2010). There are currently 45 kinesin genes identified in mammalian species, with alternative splicing expanding the total number of different kinesin proteins (Miki et al. 2001).

1.1 KINESINS ARE DEFINED BY A CHARACTERISTIC MOTOR DOMAIN

The conserved kinesin motor domain consists of ~350 amino acids and contains both the nucleotide-binding site and the microtubule-binding interface. The kinesin motor domain acts as a nucleotide-gated switch, with its conformation dependent on nucleotide status (Arnal and Wade 1998, Hirose et al. 2006, Kikkawa et al. 2001, Yun et al. 2001). The affinity of the motor domain for the microtubule depends on whether ATP, ADP·P_i, ADP or no nucleotide is bound (Hackney 1994, Ma and Taylor 1997). Therefore, the motor domain switches between high and low affinity for the microtubule as a nucleotide binds, is hydrolysed and the products are released. Interaction of the motor domain with the microtubule alters the kinetics of changes in nucleotide state and therefore affects the transitions between different microtubule affinity conformations. This reciprocal influence of nucleotide status on microtubule binding and interaction with the microtubule on nucleotide binding and hydrolysis results in a mechanochemical coupling that drives the motor activity of kinesins.

The kinesin motor domain is structurally conserved throughout the superfamily, consisting of an eight-stranded β-sheet flanked by three α-helices on either side (Kull et al. 1996, Sablin et al. 1996, Song et al. 2001). The β-strands are named β1 to β8, the helices α1 to α6, and the connecting loops, in which most of the structural diversity exists, are typically named L1 to L12.

1.1.1 THE NUCLEOTIDE-BINDING SITE

The nucleotide-binding site of kinesins is formed of four conserved motifs named N1 to N4 (Sablin et al. 1996) (Figure 1.1A and Table 1.1). These nucleotide-binding motifs contain the most highly conserved amino acid residues across the kinesin superfamily and are responsible for the binding and hydrolysis of ATP. Similar motifs are found in the myosin motor domain and in G proteins, and the topology of the nucleotide-binding site is similar among these three nucleotide-gated switch domains (Kull, Vale, and Fletterick 1998, Vale 1996, Song, Marx, and Mandelkow 2002).

The N1 motif, more commonly referred to as the p-loop, binds the α- and β-phosphates of the nucleotide. The p-loop is a common phosphate-binding motif found in proteins that bind nucleotides. Residues of the p-loop, located in L7 of the kinesin motor domain, also coordinate the Mg^{2+} ion required for tight binding of the nucleotide (Muller et al. 1999). The N2 and N3 motifs are more commonly called the 'switch motifs' by analogy with G proteins (Sack, Kull, and Mandelkow 1999). The Switch I motif is found in L9 and the Switch II motif in L11 of the kinesin motor domain (Figure 1.1A). The switch regions sense the presence or absence

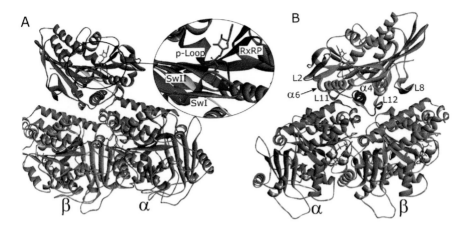

FIGURE 1.1 Structure of a Kinesin-1 motor domain in complex with the α-/β-tubulin heterodimer (pdb; 3J8Y). α-tubulin (light blue), β-tubulin (orange) and kinesin motor domain (green). (A) The conserved nucleotide-binding motifs highlighted are the p-loop (dark blue), Switch I (pink), Switch II (yellow) and RxRP (purple). The nucleotide (ATP analogue) is shown in red stick form. (B) Structural elements commonly involved in the microtubule-binding interface are highlighted as the α4 helix (dark blue), Loop 11 (pink), Loop 12 (purple), Loop 8 (yellow), Loop 2 (grey) and α6 helix (arrow).

TABLE 1.1
The Four Conserved Nucleotide-Binding Motifs Found in the Kinesin Motor Domain

Nucleotide-binding motif	N1	N2	N3	N4
Alternative names	p-loop, Walker A	Switch 1	Switch II	RxRP
Typical sequence in kinesins	GQTxxGKT	NxxSSRSH	DLAGxE	RxRP
Interacts with	α- and β-phosphate and Mg²⁺ ion	γ-phosphate	γ-phosphate	adenosine ring

of the γ-phosphate and therefore detect the nucleotide status of the motor domain. The switch loops undergo conformational changes depending on the nature of the bound nucleotide, and thereby transmit nucleotide status to the microtubule-binding interface and vice versa (Kikkawa et al. 2001, Naber et al. 2003). The N4 motif coordinates the base moiety of the nucleotide via stacking interactions with proline and the aliphatic side chains of the conserved arginine residues.

1.1.2 THE MICROTUBULE-BINDING INTERFACE

The structure of the motor domain of various kinesin families in complex with tubulin have been solved (Shang et al. 2014, Ogawa et al. 2017, Gigant et al. 2013,

Atherton et al. 2014, Wang et al. 2017, Trofimova et al. 2018, Goulet et al. 2012). Each of these structures shows a common tubulin/microtubule interface and a common orientation of this interface in relation to the α-/β-tubulin heterodimer (Figure 1.1B). The microtubule interface of the kinesin motor domain is consistently composed of the following elements of secondary structure: L8, L11/α4 and L12. The degree of involvement of other secondary structural elements (L2, α5 and α6) varies according to the kinesin family. For example, the length of L2 is considerably different between families, ranging from 2–3 residues in the Kinesin-1 family to 13–15 residues in the Kinesin-13 family. For kinesins with a long L2, this region becomes a more important element of the microtubule interface (Ogawa et al. 2004, Shipley et al. 2004, Kim, Fonseca, and Stumpff 2014).

The orientation of the microtubule-binding face of the motor domain is also preserved across kinesin families. All structures of the kinesin motor domain in complex with tubulin or the microtubule which are currently available show that the interface is centred on the α4 helix, which contacts the interface between the α- and β-tubulin subunits, the so-called 'intradimer groove' (Figure 1.1B). The L2 side of the motor domain is oriented toward the α-tubulin end of the heterodimer, whilst L8 is oriented toward the β-tubulin end. The α4 helix connects, at its N-terminus, directly to L11, which contains the Switch II nucleotide-binding motif. The length of the α4 helix has been shown to change, depending on the nucleotide- and microtubule-binding status of the motor domain (Sindelar and Downing 2010, Kikkawa et al. 2001); thus the relative lengths and conformation of L11 and α4 respond to both nucleotide and microtubule interaction status. This L11/α4 region is a particularly important route of communication between the microtubule-binding face of the motor domain and the nucleotide-binding site and is critical to the mechanochemical coupling upon which motor domain function is founded.

1.2   OUTSIDE THE MOTOR DOMAIN

Adjacent to the motor domain, on either the N- or C-terminal side, many kinesins possess a region referred to as the neck. For motile kinesins, this region is important for motor domain coordination and dictates the directionality of movement upon the microtubule (Endow and Waligora 1998, Case et al. 1997). The Kinesin-13 non-motile, microtubule-depolymerising kinesins possess a region N-terminal to the motor domain, also referred to as the neck (Maney, Wagenbach, and Wordeman 2001, Vale and Fletterick 1997), which is required for maximal activity but appears unrelated to the region of the same name in motile kinesins (Maney, Wagenbach, and Wordeman 2001, Ovechkina, Wagenbach, and Wordeman 2002).

The non-motor regions are diverse both across and within kinesin families. These regions dictate the multimerisation state of a kinesin, its cellular localisation and attachment to cargo or to adaptor proteins (Tao et al. 2006, Chu et al. 2005, Skoufias et al. 1994, Lee et al. 2010). Some kinesins even possess additional microtubule-binding sites outside the motor domain, which can increase processivity of movement

(Mayr et al. 2011) or allow crosslinking of microtubules to promote sliding (Weinger et al. 2011, Fink et al. 2009).

1.3 THE VARIOUS FUNCTIONS OF KINESINS

Kinesin superfamily members are present in all eukaryotes analysed to date (Wickstead, Gull, and Richards 2010). The kinesin motor domain can function as part of a monomer (Okada et al. 1995), dimer (Maney et al. 1998, Setou et al. 2000, Chu et al. 2005, Yildiz et al. 2004) or tetramer (Decarreau et al. 2017, Drechsler et al. 2014, Kapitein et al. 2005, Howard, Hudspeth, and Vale 1989). As a generalisation, the activity of all kinesins studied to date fall into one or both of two classes: translocating kinesins, that move directionally with respect to the microtubule, and regulating kinesins, that bind to microtubule ends and influence microtubule dynamics. Translocating kinesins may be highly processive, such that individual molecules can take many steps along a microtubule before dissociating, or less processive, able to take only one or a few steps before dissociating. Microtubule-regulating kinesins may be depolymerisers, that antagonise microtubule growth and/or promote shrinkage, polymerisers, that promote microtubule growth, or dynamics inhibitors, that antagonise both growth and disassembly of microtubules (Friel and Howard 2012).

The type of activity displayed by a kinesin is largely specified by the motor domain. Since the motor domain sequence is sufficient to place a kinesin into a family (Wickstead and Gull 2006, Lawrence et al. 2004), the location of a particular kinesin within a family gives some information as to its activity and/or function. For example, all members of the Kinesin-13 family studied to date have been found to be non-motile microtubule depolymerisers or destabilisers (Friel and Welburn 2018). Also, certain kinesin families are specialised to carry out particular functions and this affects their distribution across species. For example, several kinesin families (2, 9, 16 and 17) are cilia/flagella specific and so their expression is confined to species which contain cilia and/or flagella (Wickstead, Gull, and Richards 2010). The combination of the range of activities displayed by the various kinesin motor domains and the diversity of the non-motor regions allows kinesins to carry out a wide range of cell biological functions. These include transport of cargo such as organelles (Nangaku et al. 1994, Nakata and Hirokawa 1995, Hoepfner et al. 2005), protein complexes (Shi et al. 2004) and mRNA particles (Kanai, Dohmae, and Hirokawa 2004); intraflagellar transport particles (Kozminski, Beech, and Rosenbaum 1995) and chromosomes (Wood et al. 1997, Schaar et al. 1997); microtubule sliding (Kapitein et al. 2005, Fink et al. 2009); microtubule depolymerisation (Desai et al. 1999); and microtubule nucleation/elongation (Rome and Ohkura 2018, Hibbel et al. 2015). This range of functions means that kinesins play vital roles in many critical cellular systems, including cell division via mitosis or meiosis; development, maintenance and function of neural projections; growth, maintenance and function of cilia and flagella; and development and maintenance of cell polarity (Cross and McAinsh 2014, Goldstein and Yang 2000, Scholey 2008, Bachmann and Straube 2015, Endow, Kull, and Liu 2010, Hirokawa et al. 2009, Verhey, Dishinger, and Kee 2011).

1.4 A POTTED HISTORY OF KINESIN NOMENCLATURE AND CLASSIFICATION

The first indication of the existence of a kinesin superfamily came in the early 1990s as more and more genes were discovered that contained a ~350 amino acid region with 30–50% similarity to the motor domain of the earliest discovered kinesin (Enos and Morris 1990, Meluh and Rose 1990, Hagan and Yanagida 1990, Otsuka et al. 1991, Endow and Hatsumi 1991, McDonald and Goldstein 1990, Zhang et al. 1990, Le Guellec et al. 1991, Stewart et al. 1991, Endow, Henikoff, and Soler-Niedziela 1990, Aizawa et al. 1992, Nakagawa et al. 1997). These sequences often showed no similarity in regions outside the motor domain, suggesting that the conserved kinesin motor domain had become combined with different non-motor regions, allowing expansion of its functional capabilities. As the number of identified genes containing the kinesin motor domain rose, an increasingly complex nomenclature also grew. Some kinesins were named based on their localisation or cellular function, sometimes in advance of their identification as a kinesin, for example Ncd, non-claret disjunctional, identified in *D. melanogaster* (Walker, Salmon, and Endow 1990, McDonald, Stewart, and Goldstein 1990). Others were called KIF (an acronym for kinesin family), KLP (for kinesin-like protein) or KRP (for kinesin-related protein) and given numbers based on their chromosomal location or clone number (Aizawa et al. 1992, Stewart et al. 1991, Cole et al. 1992).

Analysis of the growing collection of motor domain sequences and the construction of phylogenetic trees began to suggest the existence of related groups within the myriad of somewhat arbitrarily named kinesin genes (Goldstein 1993, Goodson, Kang, and Endow 1994, Hirokawa 1996). This suggested that a classification system could be derived which could be used to produce a simplified nomenclature to facilitate communication between researchers in the motor protein field. Some of the first attempts at a classification of the kinesin superfamily used a nomenclature based on the position of the motor domain (Vale and Fletterick 1997, Hirokawa 1998, Miki et al. 2001). Three major groups were identified according to motor domain position within the full protein sequence, termed KIN N, KIN C and KIN I for motor domains at the N-terminus, C-terminus or Internal locations, respectively (Vale and Fletterick 1997), or KIN N, KIN C and KIN M, also referring to the motor domain located near the N-terminus, C-terminus or in the middle of the full sequence (Hirokawa 1998, Miki et al. 2001). These groups were further divided based on sequence similarity of the motor domain, eventually resulting in 11 N kinesin groups (N1–N11), two C kinesin groups (C1–C2) and one group of M kinesins (Miki et al. 2001).

As the number of identified kinesin genes continued to grow and bioinformatic techniques for performing phylogenetic analysis improved, the classification of related groups within the kinesin superfamily moved toward a consensus (Lawrence et al. 2002, Dagenbach and Endow 2004). However, the nomenclature associated with kinesin classification remained unresolved and disparate naming systems used by researchers in the field often resulted in confusion and miscommunication. Discussion among researchers in the kinesin field, and particularly following a special interest subgroup held at the 2003 meeting of the American Society for

Cell Biology, eventually led to agreement on a standardised nomenclature, with an accepted system of classification for the kinesin superfamily (Lawrence et al. 2004). Kinesins were classified into 14 families based on phylogenetic analyses, and a set of rules agreed upon as to what constituted a family and how each family should be named. Each family would bear the name 'Kinesin', an Arabic number would be used to designate individual families and recognised subfamilies would be referred to by adding a letter to the family name. It was also agreed that to gain the status of a recognised family or subfamily, a group must contain sequences from at least two kingdoms. Any sequences that could not be consistently grouped within a family would be referred to as 'orphan' kinesins. These rules, defining what constitutes a kinesin and how families should be named, remain the accepted gold standard of the kinesin field (Lawrence et al. 2004).

Whilst the rules defining distinct families and their associated nomenclature have not changed, the increase in available genome sequences from a greater diversity of organisms and advancements in phylogenetic methods have permitted more sophisticated and inclusive analyses of kinesin repertoires. In 2006, an updated kinesin phylogeny was presented, based on the analysis of complete or near-complete sets of kinesin sequences from 19 organisms spanning five of the six proposed eukaryotic supergroups (Wickstead and Gull 2006). This analysis expanded the membership of 11 of the previously defined kinesin families, identified three new families, and united two previously identified families. The Kinesin families 1–3, 5–9, 13 and 14 were again identified, and additional, previously unassigned sequences were encompassed within these families. This analysis found no support for separate Kinesin-4 and Kinesin-10 families. Instead these two families formed a single, well-supported group which was named the Kinesin-4/10 family. Increased sampling of sequences from a more diverse range of organisms resulted in the previously identified Kinesin-12 family becoming resolved into two monophyletic groups. To avoid confusion, these two distinct families were named Kinesin-15 (containing HsKIF15) and Kinesin-16 (containing HsKIF12). The only previously identified family not accounted for in this analysis was Kinesin-11; no evidence was found for the existence of this family as a monophyletic group. This analysis of an expanded set of kinesin sequences also identified a new cross-kingdom family named Kinesin-17, which consisted only of sequences from organisms that produce cilia or flagella.

An analysis of an even more comprehensive set of kinesin sequences was published in 2010 (Wickstead, Gull, and Richards 2010). This study used the kinesin repertoire from 45 organisms and ultimately an alignment of 1263 kinesin motor domain sequences to produce the phylogenetic analysis. This study confirmed the previously identified Kinesin-1, 2, 3, 4/10, 5, 6, 7, 8, 9, 13, 14, 15, 16 and 17 families and identified a further three families named Kinesin-18, 19 and 20.

The commonly accepted phylogenetic classifications of the kinesin superfamily use only the sequence of the motor domain region of the various kinesin genes analysed (Lawrence et al. 2004, Wickstead and Gull 2006, Wickstead, Gull, and Richards 2010). Therefore, the motor domain sequence alone is sufficient to place a kinesin within a family.

1.5 PURPOSE AND LAYOUT OF THIS BOOK

This book is intended to facilitate easy comparison of the individual families that comprise the kinesin superfamily. Therefore, each family for which sufficient information is currently available is allocated its own chapter and each chapter is laid out according to the same format, to allow ready comparison between families of **1) Example family members, 2) Structural information, 3) Functional properties, 4) Physiological roles** and **5) Involvement in disease** of the various members. The information available in the literature relating to the Kinesin families 7, 9 and 17–20, at the time of writing, was considered insufficient to merit complete chapters and so these families have been grouped together in a single chapter entitled 'Other Kinesins'. The nomenclature used in this work adheres to the rules laid out in Lawrence et al. (2004) and the classification is according to the analysis of Wickstead and Gull (2006) and Wickstead, Gull, and Richards (2010).

REFERENCES

Aizawa, H., Y. Sekine, R. Takemura, Z. Zhang, M. Nangaku, and N. Hirokawa. 1992. "Kinesin family in murine central nervous system." *J Cell Biol* 119(5):1287–96. doi:10.1083/jcb.119.5.1287.

Arnal, I., and R. H. Wade. 1998. "Nucleotide-dependent conformations of the kinesin dimer interacting with microtubules." *Structure* 6(1):33–8.

Atherton, J., I. Farabella, I. M. Yu, S. S. Rosenfeld, A. Houdusse, M. Topf, and C. A. Moores. 2014. "Conserved mechanisms of microtubule-stimulated ADP release, ATP binding, and force generation in transport kinesins." *Elife* 3:e03680. doi:10.7554/eLife.03680.

Bachmann, A., and A. Straube. 2015. "Kinesins in cell migration." *Biochem Soc Trans* 43(1):79–83. doi:10.1042/BST20140280.

Bloom, G. S., M. C. Wagner, K. K. Pfister, and S. T. Brady. 1988. "Native structure and physical properties of bovine brain kinesin and identification of the ATP-binding subunit polypeptide." *Biochemistry* 27(9):3409–16.

Brady, S. T. 1985. "A novel brain ATPase with properties expected for the fast axonal transport motor." *Nature* 317(6032):73–5.

Case, R. B., D. W. Pierce, N. Hom-Booher, C. L. Hart, and R. D. Vale. 1997. "The directional preference of kinesin motors is specified by an element outside of the motor catalytic domain." *Cell* 90(5):959–66.

Chu, H. M., M. Yun, D. E. Anderson, H. Sage, H. W. Park, and S. A. Endow. 2005. "Kar3 interaction with Cik1 alters motor structure and function." *EMBO J* 24(18):3214–23. doi:10.1038/sj.emboj.7600790.

Cole, D. G., W. Z. Cande, R. J. Baskin, D. A. Skoufias, C. J. Hogan, and J. M. Scholey. 1992. "Isolation of a sea urchin egg kinesin-related protein using peptide antibodies." *J Cell Sci* 101(Pt 2):291–301.

Cross, R. A., and A. McAinsh. 2014. "Prime movers: the mechanochemistry of mitotic kinesins." *Nat Rev Mol Cell Biol* 15(4):257–71. doi:10.1038/nrm3768.

Dagenbach, E. M., and S. A. Endow. 2004. "A new kinesin tree." *J Cell Sci* 117(Pt 1):3–7. doi:10.1242/jcs.00875.

Decarreau, J., M. Wagenbach, E. Lynch, A. R. Halpern, J. C. Vaughan, J. Kollman, and L. Wordeman. 2017. "The tetrameric kinesin Kif25 suppresses pre-mitotic centrosome separation to establish proper spindle orientation." *Nat Cell Biol* 19(4):384–90. doi:10.1038/ncb3486.

Desai, A., S. Verma, T. J. Mitchison, and C. E. Walczak. 1999. "Kin I kinesins are microtubule-destabilizing enzymes." *Cell* 96(1):69–78.

Drechsler, H., T. McHugh, M. R. Singleton, N. J. Carter, and A. D. McAinsh. 2014. "The Kinesin-12 Kif15 is a processive track-switching tetramer." *Elife* 3:e01724. doi:10.7554/eLife.01724.

Endow, S. A., and M. Hatsumi. 1991. "A multimember kinesin gene family in Drosophila." *Proc Natl Acad Sci U S A* 88(10):4424–7. doi:10.1073/pnas.88.10.4424.

Endow, S. A., S. Henikoff, and L. Soler-Niedziela. 1990. "Mediation of meiotic and early mitotic chromosome segregation in Drosophila by a protein related to kinesin." *Nature* 345(6270):81–3. doi:10.1038/345081a0.

Endow, S. A., F. J. Kull, and H. Liu. 2010. "Kinesins at a glance." *J Cell Sci* 123(Pt 20):3420–4. doi:10.1242/jcs.064113.

Endow, S. A., and K. W. Waligora. 1998. "Determinants of kinesin motor polarity." *Science* 281(5380):1200–2.

Enos, A. P., and N. R. Morris. 1990. "Mutation of a gene that encodes a kinesin-like protein blocks nuclear division in A. nidulans." *Cell* 60(6):1019–27.

Fink, G., L. Hajdo, K. J. Skowronek, C. Reuther, A. A. Kasprzak, and S. Diez. 2009. "The mitotic kinesin-14 Ncd drives directional microtubule-microtubule sliding." *Nat Cell Biol* 11(6):717–23. doi:10.1038/ncb1877.

Friel, C. T., and J. Howard. 2012. "Coupling of kinesin ATP turnover to translocation and microtubule regulation: one engine, many machines." *J Muscle Res Cell Motil* 33(6):377–83. doi:10.1007/s10974-012-9289-6.

Friel, C. T., and J. P. Welburn. 2018. "Parts list for a microtubule depolymerising kinesin." *Biochem Soc Trans* 46(6):1665–1672. doi:10.1042/BST20180350.

Gigant, B., W. Wang, B. Dreier, Q. Jiang, L. Pecqueur, A. Pluckthun, C. Wang, and M. Knossow. 2013. "Structure of a kinesin-tubulin complex and implications for kinesin motility." *Nat Struct Mol Biol* 20(8):1001–7. doi:10.1038/nsmb.2624.

Goldstein, L. S. 1993. "With apologies to scheherazade: tails of 1001 kinesin motors." *Annu Rev Genet* 27:319–51. doi:10.1146/annurev.ge.27.120193.001535.

Goldstein, L. S., and Z. Yang. 2000. "Microtubule-based transport systems in neurons: the roles of kinesins and dyneins." *Annu Rev Neurosci* 23:39–71. doi:10.1146/annurev.neuro.23.1.39.

Goodson, H. V., S. J. Kang, and S. A. Endow. 1994. "Molecular phylogeny of the kinesin family of microtubule motor proteins." *J Cell Sci* 107(Pt 7):1875–84.

Goulet, A., W. M. Behnke-Parks, C. V. Sindelar, J. Major, S. S. Rosenfeld, and C. A. Moores. 2012. "The structural basis of force generation by the mitotic motor kinesin-5." *J Biol Chem* 287(53):44654–66. doi:10.1074/jbc.M112.404228.

Hackney, D. D. 1994. "Evidence for alternating head catalysis by kinesin during microtubule-stimulated ATP hydrolysis." *Proc Natl Acad Sci U S A* 91(15):6865–9. doi:10.1073/pnas.91.15.6865.

Hagan, I., and M. Yanagida. 1990. "Novel potential mitotic motor protein encoded by the fission yeast cut7+ gene." *Nature* 347(6293):563–6. doi:10.1038/347563a0.

Hibbel, A., A. Bogdanova, M. Mahamdeh, A. Jannasch, M. Storch, E. Schaffer, D. Liakopoulos, and J. Howard. 2015. "Kinesin Kip2 enhances microtubule growth in vitro through length-dependent feedback on polymerization and catastrophe." *Elife* 4. doi:10.7554/eLife.10542.

Hirokawa, N. 1996. "The molecular mechanism of organelle transport along microtubules: the identification and characterization of KIFs (kinesin superfamily proteins)." *Cell Struct Funct* 21(5):357–67.

Hirokawa, N. 1998. "Kinesin and dynein superfamily proteins and the mechanism of organelle transport." *Science* 279(5350):519–26.

Hirokawa, N., Y. Noda, Y. Tanaka, and S. Niwa. 2009. "Kinesin superfamily motor proteins and intracellular transport." *Nat Rev Mol Cell Biol* 10(10):682–96. doi:10.1038/nrm2774.

Hirose, K., E. Akimaru, T. Akiba, S. A. Endow, and L. A. Amos. 2006. "Large conformational changes in a kinesin motor catalyzed by interaction with microtubules." *Mol Cell* 23(6):913–23. doi:10.1016/j.molcel.2006.07.020.

Hoepfner, S., F. Severin, A. Cabezas, B. Habermann, A. Runge, D. Gillooly, H. Stenmark, and M. Zerial. 2005. "Modulation of receptor recycling and degradation by the endosomal kinesin KIF16B." *Cell* 121(3):437–50. doi:10.1016/j.cell.2005.02.017.

Howard, J., A. J. Hudspeth, and R. D. Vale. 1989. "Movement of microtubules by single kinesin molecules." *Nature* 342(6246):154–8. doi:10.1038/342154a0.

Kanai, Y., N. Dohmae, and N. Hirokawa. 2004. "Kinesin transports RNA: isolation and characterization of an RNA-transporting granule." *Neuron* 43(4):513–25. doi:10.1016/j.neuron.2004.07.022.

Kapitein, L. C., E. J. Peterman, B. H. Kwok, J. H. Kim, T. M. Kapoor, and C. F. Schmidt. 2005. "The bipolar mitotic kinesin Eg5 moves on both microtubules that it crosslinks." *Nature* 435(7038):114–8. doi:10.1038/nature03503.

Kikkawa, M., E. P. Sablin, Y. Okada, H. Yajima, R. J. Fletterick, and N. Hirokawa. 2001. "Switch-based mechanism of kinesin motors." *Nature* 411(6836):439–45. doi:10.1038/35078000.

Kim, H., C. Fonseca, and J. Stumpff. 2014. "A unique kinesin-8 surface loop provides specificity for chromosome alignment." *Mol Biol Cell* 25(21):3319–29. doi:10.1091/mbc.E14-06-1132.

Kozminski, K. G., P. L. Beech, and J. L. Rosenbaum. 1995. "The Chlamydomonas kinesin-like protein FLA10 is involved in motility associated with the flagellar membrane." *J Cell Biol* 131(6 Pt 1):1517–27. doi:10.1083/jcb.131.6.1517.

Kull, F. J., E. P. Sablin, R. Lau, R. J. Fletterick, and R. D. Vale. 1996. "Crystal structure of the kinesin motor domain reveals a structural similarity to myosin." *Nature* 380(6574):550–5. doi:10.1038/380550a0.

Kull, F. J., R. D. Vale, and R. J. Fletterick. 1998. "The case for a common ancestor: kinesin and myosin motor proteins and G proteins." *J Muscle Res Cell Motil* 19(8):877–86.

Kuznetsov, S. A., and V. I. Gelfand. 1986. "Bovine brain kinesin is a microtubule-activated ATPase." *Proc Natl Acad Sci U S A* 83(22):8530–4. doi:10.1073/pnas.83.22.8530.

Kuznetsov, S. A., Y. A. Vaisberg, S. W. Rothwell, D. B. Murphy, and V. I. Gelfand. 1989. "Isolation of a 45-kDa fragment from the kinesin heavy chain with enhanced ATPase and microtubule-binding activities." *J Biol Chem* 264(1):589–95.

Lawrence, C. J., R. K. Dawe, K. R. Christie, D. W. Cleveland, S. C. Dawson, S. A. Endow, L. S. Goldstein, H. V. Goodson, N. Hirokawa, J. Howard, R. L. Malmberg, J. R. McIntosh, H. Miki, T. J. Mitchison, Y. Okada, A. S. Reddy, W. M. Saxton, M. Schliwa, J. M. Scholey, R. D. Vale, C. E. Walczak, and L. Wordeman. 2004. "A standardized kinesin nomenclature." *J Cell Biol* 167(1):19–22. doi:10.1083/jcb.200408113.

Lawrence, C. J., R. L. Malmberg, M. G. Muszynski, and R. K. Dawe. 2002. "Maximum likelihood methods reveal conservation of function among closely related kinesin families." *J Mol Evol* 54(1):42–53. doi:10.1007/s00239-001-0016-y.

Le Guellec, R., J. Paris, A. Couturier, C. Roghi, and M. Philippe. 1991. "Cloning by differential screening of a Xenopus cDNA that encodes a kinesin-related protein." *Mol Cell Biol* 11(6):3395–8. doi:10.1128/mcb.11.6.3395.

Lee, Y. M., E. Kim, M. Park, E. Moon, S. M. Ahn, W. Kim, K. B. Hwang, Y. K. Kim, W. Choi, and W. Kim. 2010. "Cell cycle-regulated expression and subcellular localization of a kinesin-8 member human KIF18B." *Gene* 466(1–2):16–25. doi:10.1016/j.gene.2010.06.007.

Ma, Y. Z., and E. W. Taylor. 1997. "Interacting head mechanism of microtubule-kinesin ATPase." *J Biol Chem* 272(2):724–30. doi:10.1074/jbc.272.2.724.

Maney, T., A. W. Hunter, M. Wagenbach, and L. Wordeman. 1998. "Mitotic centromere-associated kinesin is important for anaphase chromosome segregation." *J Cell Biol* 142(3):787–801. doi:10.1083/jcb.142.3.787.

Maney, T., M. Wagenbach, and L. Wordeman. 2001. "Molecular dissection of the microtubule depolymerizing activity of mitotic centromere-associated kinesin." *J Biol Chem* 276(37):34753–8. doi:10.1074/jbc.M106626200.

Mayr, M. I., M. Storch, J. Howard, and T. U. Mayer. 2011. "A non-motor microtubule binding site is essential for the high processivity and mitotic function of kinesin-8 Kif18A." *PLoS One* 6(11):e27471. doi:10.1371/journal.pone.0027471.

McDonald, H. B., and L. S. Goldstein. 1990. "Identification and characterization of a gene encoding a kinesin-like protein in Drosophila." *Cell* 61(6):991–1000.

McDonald, H. B., R. J. Stewart, and L. S. Goldstein. 1990. "The kinesin-like ncd protein of Drosophila is a minus end-directed microtubule motor." *Cell* 63(6):1159–65.

Meluh, P. B., and M. D. Rose. 1990. "KAR3, a kinesin-related gene required for yeast nuclear fusion." *Cell* 60(6):1029–41.

Miki, H., M. Setou, K. Kaneshiro, and N. Hirokawa. 2001. "All kinesin superfamily protein, KIF, genes in mouse and human." *Proc Natl Acad Sci U S A* 98(13):7004–11. doi:10.1073/pnas.111145398.

Muller, J., A. Marx, S. Sack, Y. H. Song, and E. Mandelkow. 1999. "The structure of the nucleotide-binding site of kinesin." *Biol Chem* 380(7–8):981–92. doi:10.1515/BC.1999.122.

Naber, N., S. Rice, M. Matuska, R. D. Vale, R. Cooke, and E. Pate. 2003. "EPR spectroscopy shows a microtubule-dependent conformational change in the kinesin switch 1 domain." *Biophys J* 84(5):3190–6. doi:10.1016/S0006-3495(03)70043-5.

Nakagawa, T., Y. Tanaka, E. Matsuoka, S. Kondo, Y. Okada, Y. Noda, Y. Kanai, and N. Hirokawa. 1997. "Identification and classification of 16 new kinesin superfamily (KIF) proteins in mouse genome." *Proc Natl Acad Sci U S A* 94(18):9654–9. doi:10.1073/pnas.94.18.9654.

Nakata, T., and N. Hirokawa. 1995. "Point mutation of adenosine triphosphate-binding motif generated rigor kinesin that selectively blocks anterograde lysosome membrane transport." *J Cell Biol* 131(4):1039–53. doi:10.1083/jcb.131.4.1039.

Nangaku, M., R. Sato-Yoshitake, Y. Okada, Y. Noda, R. Takemura, H. Yamazaki, and N. Hirokawa. 1994. "KIF1B, a novel microtubule plus end-directed monomeric motor protein for transport of mitochondria." *Cell* 79(7):1209–20.

Neighbors, B. W., R. C. Williams, Jr., and J. R. McIntosh. 1988. "Localization of kinesin in cultured cells." *J Cell Biol* 106(4):1193–204. doi:10.1083/jcb.106.4.1193.

Ogawa, T., R. Nitta, Y. Okada, and N. Hirokawa. 2004. "A common mechanism for microtubule destabilizers-M type kinesins stabilize curling of the protofilament using the class-specific neck and loops." *Cell* 116(4):591–602.

Ogawa, T., S. Saijo, N. Shimizu, X. Jiang, and N. Hirokawa. 2017. "Mechanism of catalytic microtubule depolymerization via KIF2-tubulin transitional conformation." *Cell Rep* 20(11):2626–38. doi:10.1016/j.celrep.2017.08.067.

Okada, Y., H. Yamazaki, Y. Sekine-Aizawa, and N. Hirokawa. 1995. "The neuron-specific kinesin superfamily protein KIF1A is a unique monomeric motor for anterograde axonal transport of synaptic vesicle precursors." *Cell* 81(5):769–80.

Otsuka, A. J., A. Jeyaprakash, J. Garcia-Anoveros, L. Z. Tang, G. Fisk, T. Hartshorne, R. Franco, and T. Born. 1991. "The C. elegans unc-104 gene encodes a putative kinesin heavy chain-like protein." *Neuron* 6(1):113–22.

Ovechkina, Y., M. Wagenbach, and L. Wordeman. 2002. "K-loop insertion restores microtubule depolymerizing activity of a 'neckless' MCAK mutant." *J Cell Biol* 159(4):557–62. doi:10.1083/jcb.200205089.

Rome, P., and H. Ohkura. 2018. "A novel microtubule nucleation pathway for meiotic spindle assembly in oocytes." *J Cell Biol* 217(10):3431–45. doi:10.1083/jcb.201803172.

Sablin, E. P., F. J. Kull, R. Cooke, R. D. Vale, and R. J. Fletterick. 1996. "Crystal structure of the motor domain of the kinesin-related motor ncd." *Nature* 380(6574):555–9. doi:10.1038/380555a0.

Sack, S., F. J. Kull, and E. Mandelkow. 1999. "Motor proteins of the kinesin family. Structures, variations, and nucleotide binding sites." *Eur J Biochem* 262(1):1–11.

Saxton, W. M., M. E. Porter, S. A. Cohn, J. M. Scholey, E. C. Raff, and J. R. McIntosh. 1988. "Drosophila kinesin: characterization of microtubule motility and ATPase." *Proc Natl Acad Sci U S A* 85(4):1109–13. doi:10.1073/pnas.85.4.1109.

Schaar, B. T., G. K. Chan, P. Maddox, E. D. Salmon, and T. J. Yen. 1997. "CENP-E function at kinetochores is essential for chromosome alignment." *J Cell Biol* 139(6):1373–82. doi:10.1083/jcb.139.6.1373.

Scholey, J. M. 2008. "Intraflagellar transport motors in cilia: moving along the cell's antenna." *J Cell Biol* 180(1):23–9. doi:10.1083/jcb.200709133.

Scholey, J. M., J. Heuser, J. T. Yang, and L. S. Goldstein. 1989. "Identification of globular mechanochemical heads of kinesin." *Nature* 338(6213):355–7. doi:10.1038/338355a0.

Scholey, J. M., M. E. Porter, P. M. Grissom, and J. R. McIntosh. 1985. "Identification of kinesin in sea urchin eggs, and evidence for its localization in the mitotic spindle." *Nature* 318(6045):483–6.

Setou, M., T. Nakagawa, D. H. Seog, and N. Hirokawa. 2000. "Kinesin superfamily motor protein KIF17 and mLin-10 in NMDA receptor-containing vesicle transport." *Science* 288(5472):1796–802.

Shang, Z., K. Zhou, C. Xu, R. Csencsits, J. C. Cochran, and C. V. Sindelar. 2014. "High-resolution structures of kinesin on microtubules provide a basis for nucleotide-gated force-generation." *Elife* 3:e04686. doi:10.7554/eLife.04686.

Shi, S. H., T. Cheng, L. Y. Jan, and Y. N. Jan. 2004. "APC and GSK-3beta are involved in mPar3 targeting to the nascent axon and establishment of neuronal polarity." *Curr Biol* 14(22):2025–32. doi:10.1016/j.cub.2004.11.009.

Shipley, K., M. Hekmat-Nejad, J. Turner, C. Moores, R. Anderson, R. Milligan, R. Sakowicz, and R. Fletterick. 2004. "Structure of a kinesin microtubule depolymerization machine." *EMBO J* 23(7):1422–32. doi:10.1038/sj.emboj.7600165.

Sindelar, C. V., and K. H. Downing. 2010. "An atomic-level mechanism for activation of the kinesin molecular motors." *Proc Natl Acad Sci U S A* 107(9):4111–6. doi:10.1073/pnas.0911208107.

Skoufias, D. A., D. G. Cole, K. P. Wedaman, and J. M. Scholey. 1994. "The carboxyl-terminal domain of kinesin heavy chain is important for membrane binding." *J Biol Chem* 269(2):1477–85.

Song, Y. H., A. Marx, and E. Mandelkow. 2002. "Structures of kinesin motor domains: implications for conformational switching involved in mechanochemical coupling." In *Molecular Motors*, edited by M. Schliwa, 287–303. Wiley-VCH Verlag GmbH & Co. KGaA.

Song, Y. H., A. Marx, J. Muller, G. Woehlke, M. Schliwa, A. Krebs, A. Hoenger, and E. Mandelkow. 2001. "Structure of a fast kinesin: implications for ATPase mechanism and interactions with microtubules." *EMBO J* 20(22):6213–25. doi:10.1093/emboj/20.22.6213.

Stewart, R. J., P. A. Pesavento, D. N. Woerpel, and L. S. Goldstein. 1991. "Identification and partial characterization of six members of the kinesin superfamily in Drosophila." *Proc Natl Acad Sci U S A* 88(19):8470–4. doi:10.1073/pnas.88.19.8470.

Tao, L., A. Mogilner, G. Civelekoglu-Scholey, R. Wollman, J. Evans, H. Stahlberg, and J. M. Scholey. 2006. "A homotetrameric kinesin-5, KLP61F, bundles microtubules and antagonizes Ncd in motility assays." *Curr Biol* 16(23):2293–302. doi:10.1016/j.cub.2006.09.064.

Trofimova, D., M. Paydar, A. Zara, L. Talje, B. H. Kwok, and J. S. Allingham. 2018. "Ternary complex of Kif2A-bound tandem tubulin heterodimers represents a kinesin-13-mediated microtubule depolymerization reaction intermediate." *Nat Commun* 9(1):2628. doi:10.1038/s41467-018-05025-7.

Vale, R. D. 1996. "Switches, latches, and amplifiers: common themes of G proteins and molecular motors." *J Cell Biol* 135(2):291–302. doi:10.1083/jcb.135.2.291.

Vale, R. D., and R. J. Fletterick. 1997. "The design plan of kinesin motors." *Annu Rev Cell Dev Biol* 13:745–77. doi:10.1146/annurev.cellbio.13.1.745.

Vale, R. D., T. S. Reese, and M. P. Sheetz. 1985. "Identification of a novel force-generating protein, kinesin, involved in microtubule-based motility." *Cell* 42(1):39–50.

Vale, R. D., B. J. Schnapp, T. S. Reese, and M. P. Sheetz. 1985a. "Movement of organelles along filaments dissociated from the axoplasm of the squid giant axon." *Cell* 40(2):449–54.

Vale, R. D., B. J. Schnapp, T. S. Reese, and M. P. Sheetz. 1985b. "Organelle, bead, and microtubule translocations promoted by soluble factors from the squid giant axon." *Cell* 40(3):559–69.

Verhey, K. J., J. Dishinger, and H. L. Kee. 2011. "Kinesin motors and primary cilia." *Biochem Soc Trans* 39(5):1120–5. doi:10.1042/BST0391120.

Walker, R. A., E. D. Salmon, and S. A. Endow. 1990. "The Drosophila claret segregation protein is a minus-end directed motor molecule." *Nature* 347(6295):780–2. doi:10.1038/347780a0.

Wang, W., S. Cantos-Fernandes, Y. Lv, H. Kuerban, S. Ahmad, C. Wang, and B. Gigant. 2017. "Insight into microtubule disassembly by kinesin-13s from the structure of Kif2C bound to tubulin." *Nat Commun* 8(1):70. doi:10.1038/s41467-017-00091-9.

Weinger, J. S., M. Qiu, G. Yang, and T. M. Kapoor. 2011. "A nonmotor microtubule binding site in kinesin-5 is required for filament crosslinking and sliding." *Curr Biol* 21(2):154–60. doi:10.1016/j.cub.2010.12.038.

Wickstead, B., and K. Gull. 2006. "A 'holistic' kinesin phylogeny reveals new kinesin families and predicts protein functions." *Mol Biol Cell* 17(4):1734–43. doi:10.1091/mbc. e05-11-1090.

Wickstead, B., K. Gull, and T. A. Richards. 2010. "Patterns of kinesin evolution reveal a complex ancestral eukaryote with a multifunctional cytoskeleton." *BMC Evol Biol* 10:110. doi:10.1186/1471-2148-10-110.

Wood, K. W., R. Sakowicz, L. S. Goldstein, and D. W. Cleveland. 1997. "CENP-E is a plus end-directed kinetochore motor required for metaphase chromosome alignment." *Cell* 91(3):357–66.

Yang, J. T., R. A. Laymon, and L. S. Goldstein. 1989. "A three-domain structure of kinesin heavy chain revealed by DNA sequence and microtubule binding analyses." *Cell* 56(5):879–89.

Yildiz, A., M. Tomishige, R. D. Vale, and P. R. Selvin. 2004. "Kinesin walks hand-over-hand." *Science* 303(5658):676–8. doi:10.1126/science.1093753.

Yun, M., X. Zhang, C. G. Park, H. W. Park, and S. A. Endow. 2001. "A structural pathway for activation of the kinesin motor ATPase." *EMBO J* 20(11):2611–8. doi:10.1093/emboj/20.11.2611.

Zhang, P., B. A. Knowles, L. S. Goldstein, and R. S. Hawley. 1990. "A kinesin-like protein required for distributive chromosome segregation in Drosophila." *Cell* 62(6):1053–62.

2 The Kinesin-1 Family
Long-Range Transporters

David D. Hackney and Alison E. Twelvetrees

CONTENTS

The Kinesin-1 family contains the founding members of the kinesin superfamily. Members of this family are general purpose motor proteins for movement towards the plus end of microtubules.

2.1 EXAMPLE FAMILY MEMBERS

Heavy Chains (HCs):

Mammalian: KIF5A, B and C
Drosophila melanogaster: KHC
Neurospora crassa: NKin
Caenorhabditis elegans: Unc116
Note: KHC is often used to refer to "kinesin heavy chain", irrespective of source. Kinesin-1 is also known as "conventional" kinesin because it was the first to be discovered and thus became the benchmark with which other kinesins are compared.

Light Chains (LCs):

Mammalian: KLC1, 2, 3 and 4
Drosophila melanogaster: KLC

2.2 STRUCTURAL INFORMATION

Kinesin-1s in higher organisms are heterotetramers composed of two heavy chains (HCs), each of which contains a motor domain, and two light chains (LCs) (see Vale and Fletterick (1997), Hirokawa et al. (2010) and Wang et al. (2015) for reviews) (Figure 2.1A). Kinesin-1 is classified as an N-terminal kinesin because the motor domains (MDs) or heads are at the N-terminus of the HCs. The HCs dimerise through the long coiled-coil regions of the stalk (Coils 1 and 2). The cargo- and LC-binding regions are located in the C-terminal region (tail). The stalk region connects the tail and head domains and serves as a flexible spacer to link the motor and cargo-binding regions.

The motor domains (MDs) consist of a core domain of approximately 340 amino acids, with sites for binding ATP and microtubules. Following the core MD on the HC is a short sequence of amino acids that is called the neck linker (NL), shown in magenta in Figures 2.1A and B. The NL in Figure 2.1B is bound to the core motor in a docked configuration that is favoured by binding of non-hydrolysable ATP analogues or transition state analogues to the motor domain, but can be undocked (disordered and not seen in X-ray structures) when ADP is bound or in the absence of a nucleotide. Reversible docking of the NL (Rice et al., 1999) plays a key role in the

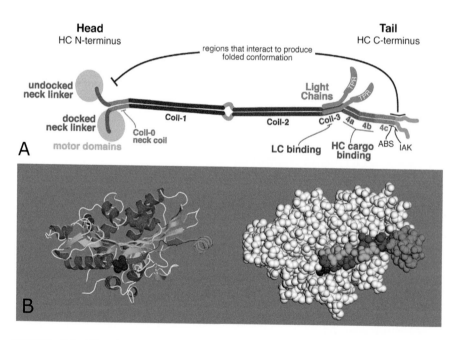

FIGURE 2.1 Kinesin-1 structure. (A) Domain organisation of Kinesin-1. Modified from ©2007 Hackney, D.D. Originally published in *J. Cell Biol.* https://doi.org/10.1083/jcb.200611082. (B) Kinesin-1 rat monomer motor domain (PDB 2kin) (Sack et al., 1997). Neck linker (NL, magenta), cover strand (cyan), helix at start of neck coil (green), critical Ile327 at start of NL (blue); and ADP in spacefill at back of view. HC: heavy chain; LC: light chain.

generation of processive movement as discussed below. Additionally, the neck linker is overlaid by the N-terminal region (shown in cyan in Figure 2.1B), designated the cover strand (Khalil et al., 2008). The NL is followed by a short coiled-coil region (Coil-0), called the neck coil (shown in green in Figures 2.1A and B), and a hinge region. The neck coil dimerises tightly (Morii et al., 1997; Tripet et al., 1997) and produces a functional dimer of heads that is sufficient for generation of processive motility. When their NLs are docked, the heads are constrained by the neck coil to be next to each other but can move away from the neck coil if undocked, as indicated by one undocked NL in Figure 2.1A. The stalk is largely coiled-coil in solution, but Coil-1 is surprisingly unstable in isolation (de Cuevas et al., 1992).

Following the stalk are two additional predicted coiled-coil regions in metazoan Kinesin-1s. The first region (Coil-3) binds the coiled-coil region of the LCs to generate the heterotetramer. The second conserved region (Coil-4) is a site for binding of cargoes directly to the HC. It was initially recognised and studied in *Neurospora* (Seiler et al., 2000), which lacks light chains and therefore all cargo must bind to the HCs. However, many HC-interacting proteins bind outside this region.

The C-terminal region of the HC binds to the MDs to form a folded conformation that is facilitated by the flexible hinge in the centre of the stalk between Coil-1 and Coil-2. The extended conformation in Figure 2.1A is only observed at high ionic strength, which weakens the tail-head interaction. Under typical physiological conditions, Kinesin-1 is in the folded state and is autoinhibited. A highly conserved IAK motif and surrounding residues in this region are required for interaction with the heads (Stock et al., 1999). Unexpectedly, only one of the tail peptides of a HC dimer binds tightly to a dimer of heads (Kaan et al., 2011). The tail also contains a nucleotide-independent microtubule-binding site (Navone et al., 1992). This positively charged auxiliary microtubule-binding site (ABS) is upstream of the IAK motif and also increases the affinity of the tails for the heads (Hackney and Stock, 2000). The extreme C-terminal region is disordered and highly variable. *Neurospora* NKin contains a conserved IAK region but lacks a conserved positively charged ABS region. *Schizosaccharomyces pombe* klp3 has a much shorter stalk and tail, with, at most, a weakly homologous IAK region.

See Wickstead et al. (2010) for a detailed phylogenetic tree of the Kinesin-1 family. Note that the designations Kinesin-1A,1B and 1C of Wickstead et al. (2010) represent an ancient divergence into three major Kinesin-1 clades and should not be confused with the more recent divergence into the three mammalian isoforms KIFA, B, C, which are all part of the Kinesin-1A clade.

The LCs contain an N-terminal coiled-coil followed by a highly charged spacer, a TPR (tetratrico peptide repeat) protein interaction domain (D'Andrea and Regan, 2003) and a C-terminal tail which is subject to extensive alternative splicing (Cyr et al., 1991; McCart et al., 2003). LCs can dimerise and interact with the HC coil-3 to produce the complete Kinesin-1 heterotetramer (Diefenbach et al., 1998). *S. pombe*, *Neurospora crassa* and *Dictyostelium discoideum* lack a gene for LCs in their genome and their HCs lack the LC-binding region of higher organisms. *Drosophila* has only one LC gene (KLC), but humans have four genes; KLC1 and KLC2 are broadly expressed, with KLC2 being more enriched in the brain (Rahman et al., 1998) and KLC3 having a specialised role in spermatids (Junco et al., 2001). Many cargoes of kinesin bind to the TPR domain, but the C-terminal tail is also involved in cargo specificity (Woźniak and Allan, 2006).

2.3 FUNCTIONAL PROPERTIES

2.3.1 MOTILITY

The principal function of Kinesin-1 is to move cargo towards the plus end of microtubules. It is exquisitely adapted to accomplish this task as demonstrated by many studies on the motile properties of the isolated protein. For these biophysical studies, Kinesin-1 has the advantage that it has a comparatively simple structure with a very compact MD that is composed of a single polypeptide chain and requires no accessory proteins for activity, unlike the other cytoskeletal motor protein families, myosin and dynein, which are much more complex. In addition, monomers and dimers of MDs from a range of species are well expressed in *Escherichia coli*, and are biochemically "well behaved" and stable if maintained with MgADP or MgATP (Hackney and McGoff, 2016). Extensive knowledge of the biophysical properties of Kinesin-1 as a motor protein has been obtained through *in vitro* reconstituted motility assays. The initial work was done with squid kinesin because of the ability to observe microtubules and kinesin-driven movement in the axoplasm from the giant axon. However, most subsequent mechanistic work has been done with *Drosophila*, mouse, rat or human Kinesin-1s, which have remarkably similar motile properties.

Assays are performed in two major configurations. One is to adsorb motors to a surface and then observe the ATP-driven movement of microtubules along the surface, often termed a gliding assay. With purified motors, casein is often used to first passivate the glass surface (Verma et al., 2008), which allows Kinesin-1s to bind, with their MDs free to interact with microtubules. A high surface density of kinesin results in many motors pushing on a microtubule in the "multi motor mode", whereas the "single motor mode" occurs at a limiting low surface density of motors, with a microtubule being pushed by only one Kinesin-1. A striking early observation was that Kinesin-1s, in the single motor mode, were capable of moving a microtubule over long distances without the microtubule diffusing away, and thus could move processively down a microtubule (Howard et al., 1989). The velocity was similar at 0.5–1 μm/s for both single and multi motor movement, however, more complex behaviour is observed when kinesin is free to move in a membrane (Grover et al., 2016). See Arpağ et al. (2019) and Belyy et al. (2016) for recent examples of how multiple and different motors on the same cargo can interact.

The other method is to attach a microtubule to a surface and watch kinesins moving along the microtubule (observed by attaching a kinesin to a bead or by attaching a fluorescent tag to a kinesin). During processive movement, Kinesin-1s track along the path of a single protofilament (Ray et al., 1993) with 8-nm steps per dimer for each ATP hydrolysed (Yildiz et al., 2004), but can take sidesteps if stalled at road blocks (Schneider et al., 2015) or if spacers are introduced between the NL and NC to allow a larger diffusional search for the tethered head (Hoeprich et al., 2014). Tracking of dimers at high temporal resolution has greatly extended our knowledge of the substeps and demonstrated that Kinesin-1 dimers rotate as they move and transmit torque to the stalk (Isojima et al., 2016; Mickolajczyk et al., 2019; Ramaiya et al., 2017). A bead with an attached Kinesin-1 can be held in an optical trap, which can be used both to track the bead and to study the role of applied force (see Greenleaf et al. (2007) for this and other methods of tracking). The force of the trap increases with the distance from the centre and can also be adjusted in a calibrated manner by varying the intensity of the trapping light. As a hindering force increases, the velocity of the kinesin decreases until movement stops at a stall force of ~7 pN/ nm in the single motor mode (Visscher et al., 1999).

2.3.2 Coupling to ATP Hydrolysis

A key feature of kinesins in general is that both the ATP-bound state and the no-nucleotide rigour state, but not the ADP state, are tightly bound to microtubules (Cross, 2016). It was, in fact, the high affinity for microtubules in the presence of the non-hydrolysable ATP analogue AMP-PNP that was the first property of kinesin to be observed (Lasek and Brady, 1985) and was used to initially purify the enzyme (Vale et al., 1985). Monomeric Kinesin-1 MDs hydrolyse ATP and release Pi rapidly, but the release of ADP, and thus the steady-state ATPase rate, is extremely slow. However, binding of the motor-ADP complex to a microtubule results in a conformational change of the motor that accelerates ADP release (Hackney, 1988). Passing through a weakly bound ADP state allows kinesin to detach from the microtubule so that it can relocate to new tubulin-binding sites further toward the plus end. This also means that a single MD cannot generate a high degree of processive movement, because when it cycles off the microtubule in the ADP state, it would diffuse away from the microtubule.

Dimers of Kinesin-1 MDs can move processively for long distances even under load without falling off the microtubule because the ATPase cycles of the two MDs are forced to be out of phase so that one or the other is always in a tight microtubule-binding state. When a dimer with one ADP per head binds to a microtubule, only one of the two ADPs is released (Hackney, 1994) to generate a tethered intermediate having one MD without a nucleotide and tightly bound to the microtubule, whereas the other MD retains its ADP as illustrated schematically in Figure 2.2. The head with the ADP can be bound to the trailing site (Figures 2.2A and C), but since the ADP state binds weakly to microtubules, it can also dissociate from the microtubule (Figures 2.2B and D), particularly at low ATP concentration (Mori et al., 2007). The NL plays a central role in coordination of the two MDs because in the two-head bound state the NLs are close to being fully extended and thus generate inter-MD tension. Changing the length of the NLs influences the coupling (Shastry and Hancock, 2011). The leading nucleotide-free MD will have an undocked NL that can be directed backwards, but the trailing head with an ADP can only bind to the micro-tubule with its NL directed forward, in order for both NLs to meet at the start of the neck coil. The trailing head cannot release the ADP rapidly because it cannot both bind to the trailing site and move its NL away from a docked orientation, as required for ADP release, without causing the neck coil to unwind. ATP binding to the lead MD will favour docking of its NL, which will force the trailing head forward to the next upstream microtubule-binding site, where it can now undock and release ADP. Hydrolysis completes the cycle.

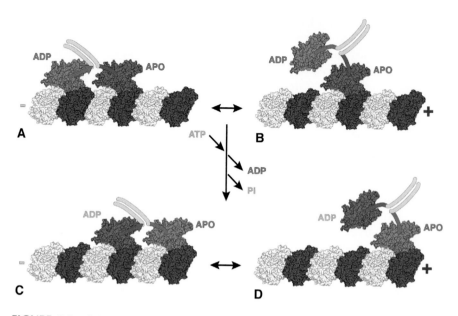

FIGURE 2.2 Schematic illustration of the interaction of Kinesin-1 with microtubules throughout its ATP turnover cycle (based on KIF1A-microtubule complex pdb 2hxh (Kikkawa and Hirokawa, 2006)). Microtubule protofilament (three dimers end to end), motor domain (nucleotide binding state indicated), neck linker (covering motor domains), helix at start of neck coil (parallel upward lines). (A) When dimeric Kinesin-1 initially binds to a microtubule ADP dissociates from only one of the two motor domains. (B) The ADP-bound motor domain dissociates from the microtubule. (C) The no nucleotide (APO) motor domain binds and hydrolyses ATP while remaining tightly associated with the microtubule. ATP hydrolysis causes the plus-end directed movement of the ADP-bound motor domain, which in the forward orientation can bind tightly to the microtubule and release ADP. (D) The trailing motor domain from (A) becomes the microtubule bound APO motor domain and the cycle starts again with dissociation from the microtubule of the new trailing ADP-bound motor domain.

The scheme depicted in Figure 2.2, however, is grossly oversimplified and is only applicable at low concentration of ATP, where hydrolysis and Pi release have time to occur before a new ATP binds to the lead apo-MD. In fact, the conformations in Figure 2.2 are referred to as the "ATP waiting state" because they will only accumulate at low ATP concentrations, where the ATP-binding rate is reduced. Considerable attention is being directed towards revealing the detailed mechanism for these transitions. One issue is whether a trailing head with ADP will remain attached to the microtubule immediately after Pi release or is so weakly bound that it dissociates even before ATP at a high concentration can bind to the leading head. At high ATP concentrations, can ATP binding start to occur at the new lead MD before hydrolysis and Pi release occurs at the new trailing head? Does the trailing head release from the microtubule in the ADP or in the ADP-Pi state? A related issue is how tightly is a MD bound to the microtubule in the ADP-Pi state. See Muretta et al. (2015), Atherton et al. (2014), Mickolajczyk et al. (2019) and Andreasson et al. (2015) for current approaches to these questions.

2.4 PHYSIOLOGICAL ROLES

Members of the Kinesin-1 family play critical roles in diverse processes, from fast axonal transport of membrane-bound organelles down axons and positioning of mitochondria and nuclei to movement of RNA granules, intermediate filaments and even dynein and microtubules (Maday et al., 2014; Hirokawa et al., 2010; Lu and Gelfand, 2017). *Drosophila* has only one HC gene, KHC, which is expressed ubiquitously and must be responsible for all Kinesin-1-driven transport. Mammals have three HC isoforms KIF5A/B/C. KIF5B is sometimes called "ubiquitous" kinesin because it is the orthologue of *Drosophila* KHC and is also expressed in all cells. KIF5A and KIF5C are enriched in the brain, with marked differences in their distribution between cell type and region (Kanai et al., 2000; Brady and Morfini, 2017). Kinesin-1 moves many different adaptors/cargoes that can bind to the LCs, HCs or both (see Gindhart (2006) for an early and now incomplete list). Since the function of Kinesin-1 is to transport cargoes, the role of this kinesin is largely defined by the identity of its cargoes and how transport is regulated.

2.4.1 CARGOES/ADAPTORS

Kinesin-1 moves both vesicular and non-vesicular cargoes, using sites on its HC or LC or both. Numerous proteins have been reported to bind to Kinesin-1 by a range of approaches, including pull-down, yeast two-hybrid and colocalisation assays, and genetic and functional interactions, with varying extents of confirmation (Gindhart, 2006; Adio et al., 2006). Their characterisation has revealed a large diversity of Kinesin-1 cargo recognition.

Many cargos bind directly to the HCs. Transmembrane cargoes can bind directly to HCs, as is the case for potassium channels (Xu et al., 2010), or through adaptor proteins, cytosolic proteins that couple a transmembrane cargo to the motor protein. Examples of adaptor proteins include GRIP1, an adaptor/scaffold to link HCs to excitatory AMPA receptors (Setou et al., 2002); HAP1, linking inhibitory $GABA_A$ receptors (Twelvetrees et al., 2010); fasciculation and elongation protein $\zeta 1$ (FEZ1 or Unc76 in *Drosophila* and *C. elegans*) (Gindhart et al., 2003; Blasius et al., 2007); and milton (TRAK1/2) to couple mitochondria through the transmembrane protein miro (Glater et al., 2006). Alternatively, cytosolic cargos linking directly to HCs without adaptors include RNA granules (Kanai et al., 2004) and intermediate filaments (Robert et al., 2019).

The LCs are just as critical as HCs for cargo binding. For example, the JIP proteins (c-Jun N-terminal kinase (JNK)-interacting protein) bind to KLCs, JNK family kinases and other proteins (Whitmarsh, 2006). Early work demonstrated the role of JIPs in the activation of kinesin and the linking of Kinesin-1 to membrane-bound cargoes, including amyloid precursor protein, APP (Matsuda et al., 2001; Scheinfeld et al., 2002; Verhey et al., 2001; Bowman et al., 2000) and the requirement, in some cases, for multiple binding partners to achieve full activation (Blasius et al., 2007; Hammond et al., 2008). Recent structural studies have provided insights into the molecular basis for their binding and activation (see Cross and Dodding (2019) and Fu and Holzbaur (2014) for reviews). JIP1 has a "W-acidic" motif near its C-terminal end, that binds to the TPR region of KLCs (Pernigo et al., 2013; Nguyen et al., 2018), whereas the nonhomologous JIP3/4 uses a coiled-coil region to bind and crosslink the TPR region (Cockburn et al., 2018). An additional aspect is that isolated LCs are autoinhibited (Yip et al., 2016). The charged spacer between the N-terminal coiled-coil region and the TPR domain contains a conserved Leucine, Phenylalanine, Proline (LFP) sequence with flanking negatively charged residues that bind to the TPRs in a region that partially overlaps with the W-acidic binding site on the TPRs and cross links these two regions. Binding of the W-acidic motif of JIP1 to the TPRs can displace the LFP region as one component in activation. Other cargoes containing W-acidic motifs include SifA-kinesin interacting protein (SKIP) (Rosa-Ferreira and Munro, 2011; Ishida et al., 2015); calsyntenin-1/alcadein-α (Konecna et al., 2006; Araki et al., 2007); Nesprin-2 (Wilson and Holzbaur, 2015); vaccinia virus protein A36R (Dodding et al., 2011); and dynein (Twelvetrees et al., 2016).

The necessity of multi-point contacts with kinesin HCs and LCs for motor recruitment is emerging as a requirement for transport. JIP1, JIP3, SKIP, HAP1 and cytoplasmic dynein all bind both HCs and LCs (Chiba et al., 2014; Fu and Holzbaur, 2013; Sanger et al., 2017; Twelvetrees et al., 2010; McGuire et al., 2006; Bowman et al., 2000; Twelvetrees et al., 2016). Binding along the stalk of Kinesin-1, not just the cargo-binding regions, may also be necessary (Fu and Holzbaur, 2013; Twelvetrees et al., 2019).

2.4.2 REGULATION

Under physiological conditions Kinesin-1 is in a folded autoinhibited conformation, produced by binding of the IAK motif of the tail of the HC to bridge the two heads and prevent microtubule-stimulated ADP release (Kaan et al., 2011). The LCs and the ABS on the HCs also interact with the heads in the folded conformation (Cai et al., 2007), but these interactions have not been structurally defined. Relief of auto-inhibition is likely to be a multistep process as discussed above.

Kinesin-1 is also regulated by post-translational modifications including phosphorylation (Brady and Morfini, 2017). When isolated from cells, Kinesin-1 contains phosphate groups on both the HC and LC (Lee and Hollenbeck, 1995). JNK3, which is upregulated in patients with Alzheimer's disease, phosphorylates MDs at Ser176, resulting in inhibition of transport (Morfini et al., 2009). Kinesin and adaptors can also be phosphorylated/dephosphorylated to influence their binding interactions, which is important for both loading and unloading of cargoes. JIP1 is phosphorylated at Ser421 by JNK with enhancement of HC binding and stimulation of anterograde movement (Fu and Holzbaur, 2013). Phosphorylation of KLC1 at Ser460 inhibits association of calsyntenin-1 (Vagnoni et al., 2011; Sobu et al., 2017). An additional mechanism of cargo unloading is binding of the heat shock protein HSP70 to the LCs (Tsai et al., 2000). HSP70 can also play a role in slow axonal transport (Terada et al., 2010). An example of transport regulation by control of the interaction between adaptor and cargo is the binding of SKIP to lysosomes, mediated by the small GTPase Arl8 (Rosa-Ferreira and Munro, 2011).

In general, microtubule-associated proteins (MAPs) such as Tau inhibit the processive movement of Kinesin-1 along microtubules (Dixit et al., 2008; Stern et al., 2017), however, MAP7 (ensconsin in *Drosophila*) actually activates Kinesin-1 (Hooikaas et al., 2019). MAP7 possesses both a domain that binds Kinesin-1 HCs in the stalk (coil-1) and a domain that binds microtubules. The additional microtubule-binding domain in the MAP7-Kinesin-1 complex increases net microtubule affinity, but the interactions are sufficiently weak and reversible that the extra microtubule binding does not exert a load on moving kinesins and slow them down.

2.5 INVOLVEMENT IN DISEASE

In recent years, many mutations in genes encoding Kinesin-1 subunits have been linked to neurological diseases in humans, connected to the dependence of neurons on long-distance transport.

Gene knockout animals gave some of the first indications that mutations in kinesin-1 genes were likely to lead to human disease. Whilst the fusion yeast *S. pombe*, which is null for Kinesin-1 HC, survives, albeit with a slow growth rate (Brazer et al., 2000), knockout or loss of function mutation of the single HC gene is lethal in complex organisms such as *Drosophila* (Brendza et al., 1999; Saxton et al., 1991), as is LC gene knockout (Gindhart et al., 1998). Similarly, homozygous disruption of the ubiquitous KIF5B in mouse is embryonic lethal (Tanaka et al., 1998), consistent with the central role of KIF5B in a broad range of cellular functions. The phenotypes of knockout mice for the neuronal kinesins KIF5A and KIF5C emphasise their key role in the nervous system; homozygous KIF5A knockout mice die shortly after birth (Xia et al., 2003), whereas homozygous KIF5C knockouts are viable, but with smaller brains and fewer motor neurons (Kanai et al., 2000). KLC1 is the most abundant of the light chains expressed in the brain and KLC1 knockout mice are smaller than their littermates, displaying pronounced motor disabilities (Rahman et al., 1999). KLC1 knockouts also exhibit age-related pathology of the retina due to defective phagocytosis in retinal pigment epithelium cells (Jiang et al., 2015).

Consistent with a key role in neuronal cell biology, mutations within genes encoding kinesin subunits cause neurological phenotypes in patients. Mutations in KIF5A have been linked to hereditary spastic paraplegia (HSP) (Reid et al., 2002) and Charcot-Marie-Tooth disease Type 2 (CMT2) (Goizet et al., 2009), both being associated with dysfunctional long axons (see Sleigh et al. (2019) for an overview). CMT2 and HSP mutations tend to occur in the MD of KIF5A and have been predicted to disrupt motility through a range of mechanisms (Ebbing et al., 2008; Jennings et al., 2017; Dutta et al., 2018). Given the uniform polarity of microtubules in the axon and the key role of KIF5A in axonal transport, it is likely that transport-deficient KIF5A shows relatively poor penetration into the distal axon. More recently, mutations in the cargo-binding regions of KIF5A have been linked to amyotrophic lateral sclerosis (ALS) (Nicolas et al., 2018; Brenner et al., 2018). The age of onset in ALS is typically later than in HSP or CMT2, but it is a much more aggressive degenerative disease, causing death two to five years after diagnosis. MD mutations in KIF5C have been linked to intellectual disability, epilepsy and malformations of cortical development (Poirier et al., 2013). Here, the differences in phenotypes linked to function-blocking mutations in KIF5A and

KIF5C MDs likely reflect the specialisations of kinesin isoforms. Characterised mutations in genes encoding LCs are currently less common, but overexpression of KLC2 has been linked to spastic paraplegia, optic atrophy and neuropathy syndrome (SPOAN) (Melo et al., 2015), whereas a mutation causing a truncated KLC4 again links Kinesin-1 to HSP (Bayrakli et al., 2015). Recently, mutation of the gene encoding a key functional partner of Kinesin-1, the adaptor JIP3, was shown to cause developmental delay and intellectual disability, and these mutations occur at sites important for the interaction with KLC2 (Cockburn et al., 2018; Platzer et al., 2019).

The broad range of patient phenotypes, even across mutations within the same Kinesin-1 subunit, highlight the need to expand our understanding of Kinesin-1 function. Conditional knockouts are currently proving useful in understanding connections between phenotypes associated with human mutations and molecular mechanisms of motor function. Conditional knockout of KIF5A in postnatal neurons causes seizures, sensory neuron degeneration and abnormal posture (Xia et al., 2003). Sensory neuron phenotypes were linked to altered neurofilament transport (Xia et al., 2003), but it's been suggested that seizures are due to impaired $GABA_A$ receptor mediated synaptic inhibition, with reduced $GABA_A$ receptors on the surface of the neurons (Nakajima et al., 2012). Some of these phenotypes are similar to that in a zebrafish mutant for KIF5A (Campbell et al., 2014). Mice with a conditional knockout of KIF5B in neurons developed hypolocomotion, motor coordination deficits and axonal transport disruption, with reduced surface expression of dopamine D2 receptors (Cromberg et al., 2019).

REFERENCES

Adio, S., J. Reth, F. Bathe, and G. Woehlke. 2006. Review: regulation mechanisms of kinesin-1. *J. Muscle Res. Cell. Motil.* 27:153–160. doi:10.1007/s10974-005-9054-1.

Andreasson, J.O.L., B. Milic, G.-Y. Chen, N.R. Guydosh, W.O. Hancock, and S.M. Block. 2015. Examining kinesin processivity within a general gating framework. *Elife*. 4. doi:10.7554/eLife.07403.

Araki, Y., T. Kawano, H. Taru, Y. Saito, S. Wada, K. Miyamoto, H. Kobayashi, H.O. Ishikawa, Y. Ohsugi, T. Yamamoto, K. Matsuno, M. Kinjo, and T. Suzuki. 2007. The novel cargo Alcadein induces vesicle association of kinesin-1 motor components and activates axonal transport. *EMBO J.* 26:1475–1486. doi:10.1038/sj.emboj.7601609.

Arpağ, G., S.R. Norris, S.I. Mousavi, V. Soppina, K.J. Verhey, W.O. Hancock, and E. Tüzel. 2019. Motor dynamics underlying cargo transport by pairs of kinesin-1 and kinesin-3 motors. *Biophys. J.* 116:1115–1126. doi:10.1016/j.bpj.2019.01.036.

Atherton, J., I. Farabella, I.-M. Yu, S.S. Rosenfeld, A. Houdusse, M. Topf, and C.A. Moores. 2014. Conserved mechanisms of microtubule-stimulated ADP release, ATP binding, and force generation in transport kinesins. *Elife*. 3:e03680. doi:10.7554/eLife.03680.

Bayrakli, F., H.G. Poyrazoglu, S. Yuksel, C. Yakicier, B. Erguner, M.S. Sagiroglu, B. Yuceturk, B. Ozer, S. Doganay, B. Tanrikulu, A. Seker, F. Akbulut, A. Ozen, H. Per, S. Kumandas, Y. Altuner Torun, M. Bayri, M. Sakar, A. Dagcinar, and I. Ziyal. 2015. Hereditary spastic paraplegia with recessive trait caused by mutation in KLC4 gene. *J. Hum. Genet.* 60:763–768. doi:10.1038/jhg.2015.109.

Belyy, V., M.A. Schlager, H. Foster, A.E. Reimer, A.P. Carter, and A. Yildiz. 2016. The mammalian dynein-dynactin complex is a strong opponent to kinesin in a tug-of-war competition. *Nat. Cell Biol.* 18:1018–1024. doi:10.1038/ncb3393.

Blasius, T.L., D. Cai, G.T. Jih, C.P. Toret, and K.J. Verhey. 2007. Two binding partners cooperate to activate the molecular motor kinesin-1. *J. Cell Biol.* 176:11–17. doi:10.1083/jcb.200605099.

Bowman, A.B., A. Kamal, B.W. Ritchings, A.V. Philp, M. McGrail, J.G. Gindhart, and L.S. Goldstein. 2000. Kinesin-dependent axonal transport is mediated by the sunday driver (SYD) protein. *Cell.* 103:583–594. doi:10.1016/s0092-8674(00)00162-8.

Brady, S.T., and G.A. Morfini. 2017. Regulation of motor proteins, axonal transport deficits and adult-onset neurodegenerative diseases. *Neurobiol. Dis.* 105:273–282. doi:10.1016/j.nbd.2017.04.010.

Brazer, S.C., H.P. Williams, T.G. Chappell, and W.Z. Cande. 2000. A fission yeast kinesin affects Golgi membrane recycling. *Yeast.* 16:149–166. doi:10.1002/(SICI)1097-0061(20000130)16:2<149::AID-YEA514>3.0.CO;2-C.

Brendza, K.M., D.J. Rose, S.P. Gilbert, and W.M. Saxton. 1999. Lethal kinesin mutations reveal amino acids important for ATPase activation and structural coupling. *J. Biol. Chem.* 274:31506–31514.

Brenner, D., R. Yilmaz, K. Müller, T. Grehl, S. Petri, T. Meyer, J. Grosskreutz, P. Weydt, W. Ruf, C. Neuwirth, M. Weber, S. Pinto, K.G. Claeys, B. Schrank, B. Jordan, A. Knehr, K. Günther, A. Hübers, D. Zeller, C. Kubisch, S. Jablonka, M. Sendtner, T. Klopstock, M. de Carvalho, A. Sperfeld, G. Borck, A.E. Volk, J. Dorst, J. Weis, M. Otto, J. Schuster, K. Del Tredici, H. Braak, K.M. Danzer, A. Freischmidt, T. Meitinger, T.M. Strom, A.C. Ludolph, P.M. Andersen, J.H. Weishaupt, and German ALS network MND-NET. 2018. Hot-spot KIF5A mutations cause familial ALS. *Brain.* 141:688–697. doi:10.1093/brain/awx370.

Cai, D., A.D. Hoppe, J.A. Swanson, and K.J. Verhey. 2007. Kinesin-1 structural organization and conformational changes revealed by FRET stoichiometry in live cells. *J. Cell Biol.* 176:51–63. doi:10.1083/jcb.200605097.

Campbell, P.D., K. Shen, M.R. Sapio, T.D. Glenn, W.S. Talbot, and F.L. Marlow. 2014. Unique function of Kinesin Kif5A in localization of mitochondria in axons. *J. Neurosci.* 34:14717–14732. doi:10.1523/JNEUROSCI.2770-14.2014.

Chiba, K., M. Araseki, K. Nozawa, K. Furukori, Y. Araki, T. Matsushima, T. Nakaya, S. Hata, Y. Saito, S. Uchida, Y. Okada, A.C. Nairn, R.J. Davis, T. Yamamoto, M. Kinjo, H. Taru, and T. Suzuki. 2014. Quantitative analysis of APP axonal transport in neurons: role of JIP1 in enhanced APP anterograde transport. *Mol. Biol. Cell.* 25:3569–3580. doi:10.1091/mbc.E14-06-1111.

Cockburn, J.J.B., S.J. Hesketh, P. Mulhair, M. Thomsen, M.J. O'Connell, and M. Way. 2018. Insights into kinesin-1 activation from the crystal structure of KLC2 bound to JIP3. *Structure.* 26:1486–1498.e6. doi:10.1016/j.str.2018.07.011.

Cromberg, L.E., T.M.M. Saez, M.G. Otero, E. Tomasella, M. Alloatti, A. Damianich, V. Pozo Devoto, J. Ferrario, D. Gelman, M. Rubinstein, and T.L. Falzone. 2019. Neuronal KIF5b deletion induces striatum-dependent locomotor impairments and defects in membrane presentation of dopamine D2 receptors. *J. Neurochem.* 149:362–380. doi:10.1111/jnc.14665.

Cross, J.A., and M.P. Dodding. 2019. Motor-cargo adaptors at the organelle-cytoskeleton interface. *Curr. Opin. Cell Biol.* 59:16–23. doi:10.1016/j.ceb.2019.02.010.

Cross, R.A. 2016. Review: mechanochemistry of the kinesin-1 ATPase. *Biopolymers.* 105:476–482. doi:10.1002/bip.22862.

de Cuevas, M., T. Tao, and L.S. Goldstein. 1992. Evidence that the stalk of Drosophila kinesin heavy chain is an alpha-helical coiled coil. *J. Cell Biol.* 116:957–965. doi:10.1083/jcb.116.4.957.

Cyr, J.L., K.K. Pfister, G.S. Bloom, C.A. Slaughter, and S.T. Brady. 1991. Molecular genetics of kinesin light chains: generation of isoforms by alternative splicing. *Proc. Natl. Acad. Sci. U.S.A.* 88:10114–10118. doi:10.1073/pnas.88.22.10114.

D'Andrea, L.D., and L. Regan. 2003. TPR proteins: the versatile helix. *Trends Biochem. Sci.* 28:655–662. doi:10.1016/j.tibs.2003.10.007.

Diefenbach, R.J., J.P. Mackay, P.J. Armati, and A.L. Cunningham. 1998. The C-terminal region of the stalk domain of ubiquitous human kinesin heavy chain contains the binding site for kinesin light chain. *Biochemistry.* 37:16663–16670. doi:10.1021/bi981163r.

Dixit, R., J.L. Ross, Y.E. Goldman, and E.L.F. Holzbaur. 2008. Differential regulation of dynein and kinesin motor proteins by tau. *Science.* 319:1086–1089. doi:10.1126/science.1152993.

Dodding, M.P., R. Mitter, A.C. Humphries, and M. Way. 2011. A kinesin-1 binding motif in vaccinia virus that is widespread throughout the human genome. *EMBO J.* 30:4523–4538. doi:10.1038/emboj.2011.326.

Dutta, M., M.R. Diehl, J.N. Onuchic, and B. Jana. 2018. Structural consequences of hereditary spastic paraplegia disease-related mutations in kinesin. *Proc. Natl. Acad. Sci. U.S.A.* 115:E10822–E10829. doi:10.1073/pnas.1810622115.

Ebbing, B., K. Mann, A. Starosta, J. Jaud, L. Schöls, R. Schüle, and G. Woehlke. 2008. Effect of spastic paraplegia mutations in KIF5A kinesin on transport activity. *Hum. Mol. Genet.* 17:1245–1252. doi:10.1093/hmg/ddn014.

Fu, M., and E.L.F. Holzbaur. 2013. JIP1 regulates the directionality of APP axonal transport by coordinating kinesin and dynein motors. *J. Cell Biol.* 202:495–508. doi:10.1083/jcb.201302078.

Fu, M., and E.L.F. Holzbaur. 2014. Integrated regulation of motor-driven organelle transport by scaffolding proteins. *Trends Cell Biol.* 24:564–574. doi:10.1016/j.tcb.2014.05.002.

Gindhart, J.G. 2006. Towards an understanding of kinesin-1 dependent transport pathways through the study of protein-protein interactions. *Brief. Funct. Genomic Proteomic.* 5:74–86. doi:10.1093/bfgp/ell002.

Gindhart, J.G., J. Chen, M. Faulkner, R. Gandhi, K. Doerner, T. Wisniewski, and A. Nandlestadt. 2003. The kinesin-associated protein UNC-76 is required for axonal transport in the Drosophila nervous system. *Mol. Biol. Cell.* 14:3356–3365. doi:10.1091/mbc.e02-12-0800.

Gindhart, J.G., C.J. Desai, S. Beushausen, K. Zinn, and L.S. Goldstein. 1998. Kinesin light chains are essential for axonal transport in Drosophila. *J. Cell Biol.* 141:443–454.

Glater, E.E., L.J. Megeath, R.S. Stowers, and T.L. Schwarz. 2006. Axonal transport of mitochondria requires milton to recruit kinesin heavy chain and is light chain independent. *J. Cell Biol.* 173:545–557. doi:10.1083/jcb.200601067.

Goizet, C., A. Boukhris, E. Mundwiller, C. Tallaksen, S. Forlani, A. Toutain, N. Carriere, V. Paquis, C. Depienne, A. Durr, G. Stevanin, and A. Brice. 2009. Complicated forms of autosomal dominant hereditary spastic paraplegia are frequent in SPG10. *Hum. Mutat.* 30:E376–E385. doi:10.1002/humu.20920.

Greenleaf, W.J., M.T. Woodside, and S.M. Block. 2007. High-resolution, single-molecule measurements of biomolecular motion. *Annu. Rev. Biophys. Biomol. Struct.* 36:171–190. doi:10.1146/annurev.biophys.36.101106.101451.

Grover, R., J. Fischer, F.W. Schwarz, W.J. Walter, P. Schwille, and S. Diez. 2016. Transport efficiency of membrane-anchored kinesin-1 motors depends on motor density and diffusivity. *Proc. Natl. Acad. Sci. U.S.A.* 113:E7185–E7193. doi:10.1073/pnas.1611398113.

Hackney, D.D. 1988. Kinesin ATPase: rate-limiting ADP release. *Proc. Natl. Acad. Sci. U.S.A.* 85:6314–6318.

Hackney, D.D. 1994. Evidence for alternating head catalysis by kinesin during microtubule-stimulated ATP hydrolysis. *Proc. Natl. Acad. Sci. U.S.A.* 91:6865–6869.

Hackney, D.D. 2007. Jump-starting kinesin. *J. Cell Biol.* 176:7–9. doi:10.1083/jcb.200611082.

Hackney, D.D., and M.S. McGoff. 2016. Nucleotide-free kinesin motor domains reversibly convert to an inactive conformation with characteristics of a molten globule. *Arch. Biochem. Biophys.* 608:42–51. doi:10.1016/j.abb.2016.08.019.

Hackney, D.D., and M.F. Stock. 2000. Kinesin's IAK tail domain inhibits initial microtubule-stimulated ADP release. *Nat. Cell Biol.* 2:257–260. doi:10.1038/35010525.

Hammond, J.W., K. Griffin, G.T. Jih, J. Stuckey, and K.J. Verhey. 2008. Co-operative versus independent transport of different cargoes by Kinesin-1. *Traffic.* 9:725–741. doi:10.1111/j.1600-0854.2008.00722.x.

Hirokawa, N., S. Niwa, and Y. Tanaka. 2010. Molecular motors in neurons: transport mechanisms and roles in brain function, development, and disease. *Neuron.* 68:610–638. doi:10.1016/j.neuron.2010.09.039.

Hoeprich, G.J., A.R. Thompson, D.P. McVicker, W.O. Hancock, and C.L. Berger. 2014. Kinesin's neck-linker determines its ability to navigate obstacles on the microtubule surface. *Biophys. J.* 106:1691–1700. doi:10.1016/j.bpj.2014.02.034.

Hooikaas, P.J., M. Martin, T. Mühlethaler, G.-J. Kuijntjes, C.A.E. Peeters, E.A. Katrukha, L. Ferrari, R. Stucchi, D.G.F. Verhagen, W.E. van Riel, I. Grigoriev, A.F.M. Altelaar, C.C. Hoogenraad, S.G.D. Rüdiger, M.O. Steinmetz, L.C. Kapitein, and A. Akhmanova. 2019. MAP7 family proteins regulate kinesin-1 recruitment and activation. *J. Cell Biol.* 218:1298–1318. doi:10.1083/jcb.201808065.

Howard, J., A.J. Hudspeth, and R.D. Vale. 1989. Movement of microtubules by single kinesin molecules. *Nature.* 342:154–158. doi:10.1038/342154a0.

Ishida, M., N. Ohbayashi, and M. Fukuda. 2015. Rab1A regulates anterograde melanosome transport by recruiting kinesin-1 to melanosomes through interaction with SKIP. *Sci. Rep.* 5:8238. doi:10.1038/srep08238.

Isojima, H., R. Iino, Y. Niitani, H. Noji, and M. Tomishige. 2016. Direct observation of intermediate states during the stepping motion of kinesin-1. *Nat. Chem. Biol.* 12:290–297. doi:10.1038/nchembio.2028.

Jennings, S., M. Chenevert, L. Liu, M. Mottamal, E.J. Wojcik, and T.M. Huckaba. 2017. Characterization of kinesin switch I mutations that cause hereditary spastic paraplegia. *PLoS ONE.* 12:e0180353. doi:10.1371/journal.pone.0180353.

Jiang, M., J. Esteve-Rudd, V.S. Lopes, T. Diemer, C. Lillo, A. Rump, and D.S. Williams. 2015. Microtubule motors transport phagosomes in the RPE, and lack of KLC1 leads to AMD-like pathogenesis. *J. Cell Biol.* 210:595–611. doi:10.1083/jcb.201410112.

Junco, A., B. Bhullar, H.A. Tarnasky, and F.A. van der Hoorn. 2001. Kinesin light-chain KLC3 expression in testis is restricted to spermatids. *Biol. Reprod.* 64:1320–1330. doi:10.1095/biolreprod64.5.1320.

Kaan, H.Y.K., D.D. Hackney, and F. Kozielski. 2011. The structure of the kinesin-1 motor-tail complex reveals the mechanism of autoinhibition. *Science.* 333:883–885. doi:10.1126/science.1204824.

Kanai, Y., N. Dohmae, and N. Hirokawa. 2004. Kinesin transports RNA: isolation and characterization of an RNA-transporting granule. *Neuron.* 43:513–525. doi:10.1016/j.neuron.2004.07.022.

Kanai, Y., Y. Okada, Y. Tanaka, A. Harada, S. Terada, and N. Hirokawa. 2000. KIF5C, a novel neuronal kinesin enriched in motor neurons. *J. Neurosci.* 20:6374–6384.

Khalil, A.S., D.C. Appleyard, A.K. Labno, A. Georges, M. Karplus, A.M. Belcher, W. Hwang, and M.J. Lang. 2008. Kinesin's cover-neck bundle folds forward to generate force. *Proc. Natl. Acad. Sci. U.S.A.* 105:19247–19252. doi:10.1073/pnas.0805147105.

Kikkawa, M., and N. Hirokawa. 2006. High-resolution cryo-EM maps show the nucleotide binding pocket of KIF1A in open and closed conformations. *EMBO J.* 25:4187–4194. doi:10.1038/sj.emboj.7601299.

Konecna, A., R. Frischknecht, J. Kinter, A. Ludwig, M. Steuble, V. Meskenaite, M. Indermühle, M. Engel, C. Cen, J.-M. Mateos, P. Streit, and P. Sonderegger. 2006. Calsyntenin-1 docks vesicular cargo to kinesin-1. *Mol. Biol. Cell.* 17:3651–3663. doi:10.1091/mbc.e06-02-0112.

Lasek, R.J., and S.T. Brady. 1985. Attachment of transported vesicles to microtubules in axoplasm is facilitated by AMP-PNP. *Nature.* 316:645–647. doi:10.1038/316645a0.

Lee, K.D., and P.J. Hollenbeck. 1995. Phosphorylation of kinesin in vivo correlates with organelle association and neurite outgrowth. *J. Biol. Chem.* 270:5600–5605. doi:10.1074/jbc.270.10.5600.

Lu, W., and V.I. Gelfand. 2017. Moonlighting motors: kinesin, dynein, and cell polarity. *Trends Cell Biol.* 27:505–514. doi:10.1016/j.tcb.2017.02.005.

Maday, S., A.E. Twelvetrees, A.J. Moughamian, and E.L.F. Holzbaur. 2014. Axonal transport: cargo-specific mechanisms of motility and regulation. *Neuron.* 84:292–309. doi:10.1016/j.neuron.2014.10.019.

Matsuda, S., T. Yasukawa, Y. Homma, Y. Ito, T. Niikura, T. Hiraki, S. Hirai, S. Ohno, Y. Kita, M. Kawasumi, K. Kouyama, T. Yamamoto, J.M. Kyriakis, and I. Nishimoto. 2001. c-Jun N-terminal kinase (JNK)-interacting protein-1b/islet-brain-1 scaffolds Alzheimer's amyloid precursor protein with JNK. *J. Neurosci.* 21:6597–6607.

McCart, A.E., D. Mahony, and J.A. Rothnagel. 2003. Alternatively spliced products of the human kinesin light chain 1 (KNS2) gene. *Traffic.* 4:576–580.

McGuire, J.R., J. Rong, S.-H. Li, and X.-J. Li. 2006. Interaction of huntingtin-associated protein-1 with kinesin light chain: implications in intracellular trafficking in neurons. *J. Biol. Chem.* 281:3552–3559. doi:10.1074/jbc.M509806200.

Melo, U.S., L.I. Macedo-Souza, T. Figueiredo, A.R. Muotri, J.G. Gleeson, G. Coux, P. Armas, N.B. Calcaterra, J.P. Kitajima, S. Amorim, T.R. Olávio, K. Griesi-Oliveira, G.C. Coatti, C.R.R. Rocha, M. Martins-Pinheiro, C.F.M. Menck, M.S. Zaki, F. Kok, M. Zatz, and S. Santos. 2015. Overexpression of KLC2 due to a homozygous deletion in the non-coding region causes SPOAN syndrome. *Hum. Mol. Genet.* 24:6877–6885. doi:10.1093/hmg/ddv388.

Mickolajczyk, K.J., A.S.I. Cook, J.P. Jevtha, J. Fricks, and W.O. Hancock. 2019. Insights into kinesin-1 stepping from simulations and tracking of gold nanoparticle-labeled motors. *Biophys. J.* doi:10.1016/j.bpj.2019.06.010.

Morfini, G.A., Y.-M. You, S.L. Pollema, A. Kaminska, K. Liu, K. Yoshioka, B. Björkblom, E.T. Coffey, C. Bagnato, D. Han, C.-F. Huang, G. Banker, G. Pigino, and S.T. Brady. 2009. Pathogenic huntingtin inhibits fast axonal transport by activating JNK3 and phosphorylating kinesin. *Nat. Neurosci.* 12:864–871. doi:10.1038/nn.2346.

Mori, T., R.D. Vale, and M. Tomishige. 2007. How kinesin waits between steps. *Nature.* 450:750–754. doi:10.1038/nature06346.

Morii, H., T. Takenawa, F. Arisaka, and T. Shimizu. 1997. Identification of kinesin neck region as a stable alpha-helical coiled coil and its thermodynamic characterization. *Biochemistry.* 36:1933–1942. doi:10.1021/bi9623921.

Muretta, J.M., Y. Jun, S.P. Gross, J. Major, D.D. Thomas, and S.S. Rosenfeld. 2015. The structural kinetics of switch-1 and the neck linker explain the functions of kinesin-1 and Eg5. *Proc. Natl. Acad. Sci. U.S.A.* 112:E6606–E6613. doi:10.1073/pnas.1512305112.

Nakajima, K., X. Yin, Y. Takei, D.-H. Seog, N. Homma, and N. Hirokawa. 2012. Molecular motor KIF5A is essential for GABA(A) receptor transport, and KIF5A deletion causes epilepsy. *Neuron.* 76:945–961. doi:10.1016/j.neuron.2012.10.012.

Navone, F., J. Niclas, N. Hom-Booher, L. Sparks, H.D. Bernstein, G. McCaffrey, and R.D. Vale. 1992. Cloning and expression of a human kinesin heavy chain gene: interaction of the COOH-terminal domain with cytoplasmic microtubules in transfected CV-1 cells. *J. Cell Biol.* 117:1263–1275. doi:10.1083/jcb.117.6.1263.

Nguyen, T.Q., M. Aumont-Nicaise, J. Andreani, C. Velours, M. Chenon, F. Vilela, C. Geneste, P.F. Varela, P. Llinas, and J. Ménétrey. 2018. Characterization of the binding mode of JNK-interacting protein 1 (JIP1) to kinesin-light chain 1 (KLC1). *J. Biol. Chem.* 293:13946–13960. doi:10.1074/jbc.RA118.003916.

Nicolas, A., K.P. Kenna, A.E. Renton, N. Ticozzi, F. Faghri, R. Chia, J.A. Dominov, B.J. Kenna, M.A. Nalls, P. Keagle, A.M. Rivera, W. van Rheenen, N.A. Murphy, J.J.F.A. van Vugt, J.T. Geiger, R.A. Van der Spek, H.A. Pliner, Shankaracharya, B.N. Smith, G. Marangi, S.D. Topp, Y. Abramzon, A.S. Gkazi, J.D. Eicher, A. Kenna, F.O. Logullo, I.L. Simone,

G. Logroscino, F. Salvi, I. Bartolomei, G. Borghero, M.R. Murru, E. Costantino, C. Pani, R. Puddu, C. Caredda, V. Piras, S. Tranquilli, S. Cuccu, D. Corongiu, M. Melis, A. Milia, F. Marrosu, M.G. Marrosu, G. Floris, A. Cannas, M. Capasso, C. Caponnetto, G. Mancardi, P. Origone, P. Mandich, F.L. Conforti, S. Cavallaro, G. Mora, K. Marinou, R. Sideri, S. Penco, L. Mosca, C. Lunetta, G.L. Pinter, M. Corbo, N. Riva, P. Carrera, P. Volanti, J. Mandrioli, N. Fini, A. Fasano, L. Tremolizzo, A. Arosio, C. Ferrarese, F. Trojsi, G. Tedeschi, M.R. Monsurrò, G. Piccirillo, C. Femiano, A. Ticca, E. Ortu, V. La Bella, R. Spataro, T. Colletti, M. Sabatelli, M. Zollino, A. Conte, M. Luigetti, S. Lattante, M. Santarelli, A. Petrucci, M. Pugliatti, A. Pirisi, L.D. Parish, P. Occhineri, F. Giannini, S. Battistini, C. Ricci, M. Benigni, T.B. Cau, D. Loi, A. Calvo, et al. 2018. Genome-wide analyses identify KIF5A as a novel ALS gene. *Neuron.* 97:1268–1283. e6. doi:10.1016/j.neuron.2018.02.027.

Pernigo, S., A. Lamprecht, R.A. Steiner, and M.P. Dodding. 2013. Structural basis for kinesin-1: cargo recognition. *Science.* 340:356–359. doi:10.1126/science.1234264.

Platzer, K., H. Sticht, S.L. Edwards, W. Allen, K.M. Angione, M.T. Bonati, C. Brasington, M.T. Cho, L.A. Demmer, T. Falik-Zaccai, C.N. Gamble, Y. Hellenbroich, M. Iascone, F. Kok, S. Mahida, H. Mandel, T. Marquardt, K. McWalter, B. Panis, A. Pepler, H. Pinz, L. Ramos, D.N. Shinde, C. Smith-Hicks, A.P.A. Stegmann, P. Stöbe, C.T.R.M. Stumpel, C. Wilson, J.R. Lemke, N. Di Donato, K.G. Miller, and R. Jamra. 2019. De novo variants in MAPK8IP3 cause intellectual disability with variable brain anomalies. *Am. J. Hum. Genet.* 104:203–212. doi:10.1016/j.ajhg.2018.12.008.

Poirier, K., N. Lebrun, L. Broix, G. Tian, Y. Saillour, C. Boscheron, E. Parrini, S. Valence, B.S. Pierre, M. Oger, D. Lacombe, D. Geneviève, E. Fontana, F. Darra, C. Cances, M. Barth, D. Bonneau, B.D. Bernadina, S. N'guyen, C. Gitiaux, P. Parent, V. des Portes, J.M. Pedespan, V. Legrez, L. Castelnau-Ptakine, P. Nitschke, T. Hieu, C. Masson, D. Zelenika, A. Andrieux, F. Francis, R. Guerrini, N.J. Cowan, N. Bahi-Buisson, and J. Chelly. 2013. Mutations in TUBG1, DYNC1H1, KIF5C and KIF2A cause malformations of cortical development and microcephaly. *Nat. Genet.* 45:639–647. doi:10.1038/ng.2613.

Rahman, A., D.S. Friedman, and L.S. Goldstein. 1998. Two kinesin light chain genes in mice. Identification and characterization of the encoded proteins. *J. Biol. Chem.* 273:15395–15403. doi:10.1074/jbc.273.25.15395.

Rahman, A., A. Kamal, E.A. Roberts, and L.S. Goldstein. 1999. Defective kinesin heavy chain behavior in mouse kinesin light chain mutants. *J. Cell Biol.* 146:1277–1288. doi:10.1083/jcb.146.6.1277.

Ramaiya, A., B. Roy, M. Bugiel, and E. Schäffer. 2017. Kinesin rotates unidirectionally and generates torque while walking on microtubules. *Proc. Natl. Acad. Sci. U.S.A.* 114:10894–10899. doi:10.1073/pnas.1706985114.

Ray, S., E. Meyhöfer, R.A. Milligan, and J. Howard. 1993. Kinesin follows the microtubule's protofilament axis. *J. Cell Biol.* 121:1083–1093. doi:10.1083/jcb.121.5.1083.

Reid, E., M. Kloos, A. Ashley-Koch, L. Hughes, S. Bevan, I.K. Svenson, F.L. Graham, P.C. Gaskell, A. Dearlove, M.A. Pericak-Vance, D.C. Rubinsztein, and D.A. Marchuk. 2002. A kinesin heavy chain (KIF5A) mutation in hereditary spastic paraplegia (SPG10). *Am. J. Hum. Genet.* 71:1189–1194. doi:10.1086/344210.

Rice, S., A.W. Lin, D. Safer, C.L. Hart, N. Naber, B.O. Carragher, S.M. Cain, E. Pechatnikova, E.M. Wilson-Kubalek, M. Whittaker, E. Pate, R. Cooke, E.W. Taylor, R.A. Milligan, and R.D. Vale. 1999. A structural change in the kinesin motor protein that drives motility. *Nature.* 402:778–784. doi:10.1038/45483.

Robert, A., P. Tian, S.A. Adam, M. Kittisopikul, K. Jaqaman, R.D. Goldman, and V.I. Gelfand. 2019. Kinesin-dependent transport of keratin filaments: a unified mechanism for intermediate filament transport. *FASEB J.* 33:388–399. doi:10.1096/fj.201800604R.

Rosa-Ferreira, C., and S. Munro. 2011. Arl8 and SKIP act together to link lysosomes to kinesin-1. *Dev. Cell.* 21:1171–1178. doi:10.1016/j.devcel.2011.10.007.

Sack, S., J. Müller, A. Marx, M. Thormählen, E.M. Mandelkow, S.T. Brady, and E. Mandelkow. 1997. X-ray structure of motor and neck domains from rat brain kinesin. *Biochemistry.* 36:16155–16165. doi:10.1021/bi9722498.

Sanger, A., Y.Y. Yip, T.S. Randall, S. Pernigo, R.A. Steiner, and M.P. Dodding. 2017. SKIP controls lysosome positioning using a composite kinesin-1 heavy and light chain-binding domain. *J. Cell. Sci.* 130:1637–1651. doi:10.1242/jcs.198267.

Saxton, W.M., J. Hicks, L.S. Goldstein, and E.C. Raff. 1991. Kinesin heavy chain is essential for viability and neuromuscular functions in Drosophila, but mutants show no defects in mitosis. *Cell.* 64:1093–1102.

Scheinfeld, M.H., R. Roncarati, P. Vito, P.A. Lopez, M. Abdallah, and L. D'Adamio. 2002. Jun NH2-terminal kinase (JNK) interacting protein 1 (JIP1) binds the cytoplasmic domain of the Alzheimer's beta-amyloid precursor protein (APP). *J. Biol. Chem.* 277:3767–3775. doi:10.1074/jbc.M108357200.

Schneider, R., T. Korten, W.J. Walter, and S. Diez. 2015. Kinesin-1 motors can circumvent permanent roadblocks by side-shifting to neighboring protofilaments. *Biophys. J.* 108:2249–2257. doi:10.1016/j.bpj.2015.03.048.

Seiler, S., J. Kirchner, C. Horn, A. Kallipolitou, G. Woehlke, and M. Schliwa. 2000. Cargo binding and regulatory sites in the tail of fungal conventional kinesin. *Nat. Cell Biol.* 2:333–338. doi:10.1038/35014022.

Setou, M., D.-H. Seog, Y. Tanaka, Y. Kanai, Y. Takei, M. Kawagishi, and N. Hirokawa. 2002. Glutamate-receptor-interacting protein GRIP1 directly steers kinesin to dendrites. *Nature.* 417:83–87. doi:10.1038/nature743.

Shastry, S., and W.O. Hancock. 2011. Interhead tension determines processivity across diverse N-terminal kinesins. *Proc. Natl. Acad. Sci. U.S.A.* 108:16253–16258. doi:10.1073/pnas.1102628108.

Sleigh, J.N., A.M. Rossor, A.D. Fellows, A.P. Tosolini, and G. Schiavo. 2019. Axonal transport and neurological disease. *Nat. Rev. Neurol.* 15:691–703. doi:10.1038/s41582-019-0257-2.

Sobu, Y., K. Furukori, K. Chiba, A.C. Nairn, M. Kinjo, S. Hata, and T. Suzuki. 2017. Phosphorylation of multiple sites within an acidic region of Alcadein α is required for kinesin-1 association and Golgi exit of Alcadein α cargo. *Mol. Biol. Cell.* 28:3844–3856. doi:10.1091/mbc.E17-05-0301.

Stern, J.L., D.V. Lessard, G.J. Hoeprich, G.A. Morfini, and C.L. Berger. 2017. Phosphoregulation of Tau modulates inhibition of kinesin-1 motility. *Mol. Biol. Cell.* 28:1079–1087. doi:10.1091/mbc.E16-10-0728.

Stock, M.F., J. Guerrero, B. Cobb, C.T. Eggers, T.G. Huang, X. Li, and D.D. Hackney. 1999. Formation of the compact confomer of kinesin requires a COOH-terminal heavy chain domain and inhibits microtubule-stimulated ATPase activity. *J. Biol. Chem.* 274:14617–14623.

Tanaka, Y., Y. Kanai, Y. Okada, S. Nonaka, S. Takeda, A. Harada, and N. Hirokawa. 1998. Targeted disruption of mouse conventional kinesin heavy chain kif5B, results in abnormal perinuclear clustering of mitochondria. *Cell.* 93:1147–1158. doi:10.1016/S0092-8674(00)81459-2.

Terada, S., M. Kinjo, M. Aihara, Y. Takei, and N. Hirokawa. 2010. Kinesin-1/Hsc70-dependent mechanism of slow axonal transport and its relation to fast axonal transport. *EMBO J.* 29:843–854. doi:10.1038/emboj.2009.389.

Tripet, B., R.D. Vale, and R.S. Hodges. 1997. Demonstration of coiled-coil interactions within the kinesin neck region using synthetic peptides. Implications for motor activity. *J. Biol. Chem.* 272:8946–8956. doi:10.1074/jbc.272.14.8946.

Tsai, M.Y., G. Morfini, G. Szebenyi, and S.T. Brady. 2000. Release of kinesin from vesicles by hsc70 and regulation of fast axonal transport. *Mol. Biol. Cell.* 11:2161–2173. doi:10.1091/mbc.11.6.2161.

Twelvetrees, A.E., F. Lesept, E.L.F. Holzbaur, and J.T. Kittler. 2019. The adaptor proteins HAP1a and GRIP1 collaborate to activate kinesin-1 isoform KIF5C. *J. Cell. Sci.* doi:10.1242/jcs.215822.

Twelvetrees, A.E., S. Pernigo, A. Sanger, P. Guedes-Dias, G. Schiavo, R.A. Steiner, M.P. Dodding, and E.L.F. Holzbaur. 2016. The dynamic localization of cytoplasmic dynein in neurons is driven by kinesin-1. *Neuron.* 90:1000–1015. doi:10.1016/j.neuron.2016.04.046.

Twelvetrees, A.E., E.Y. Yuen, I.L. Arancibia-Carcamo, A.F. MacAskill, P. Rostaing, M.J. Lumb, S. Humbert, A. Triller, F. Saudou, Z. Yan, and J.T. Kittler. 2010. Delivery of GABAARs to synapses is mediated by HAP1-KIF5 and disrupted by mutant huntingtin. *Neuron.* 65:53–65. doi:10.1016/j.neuron.2009.12.007.

Vagnoni, A., L. Rodriguez, C. Manser, K.J. De Vos, and C.C.J. Miller. 2011. Phosphorylation of kinesin light chain 1 at serine 460 modulates binding and trafficking of calsyntenin-1. *J. Cell. Sci.* 124:1032–1042. doi:10.1242/jcs.075168.

Vale, R.D., and R.J. Fletterick. 1997. The design plan of kinesin motors. *Annu. Rev. Cell Dev. Biol.* 13:745–777. doi:10.1146/annurev.cellbio.13.1.745.

Vale, R.D., T.S. Reese, and M.P. Sheetz. 1985. Identification of a novel force-generating protein, kinesin, involved in microtubule-based motility. *Cell.* 42:39–50. doi:10.1016/s0092-8674(85)80099-4.

Verhey, K.J., D. Meyer, R. Deehan, J. Blenis, B.J. Schnapp, T.A. Rapoport, and B. Margolis. 2001. Cargo of kinesin identified as JIP scaffolding proteins and associated signaling molecules. *J. Cell Biol.* 152:959–970. doi:10.1083/jcb.152.5.959.

Verma, V., W.O. Hancock, and J.M. Catchmark. 2008. The role of casein in supporting the operation of surface bound kinesin. *J Biol Eng.* 2:14. doi:10.1186/1754-1611-2-14.

Visscher, K., M.J. Schnitzer, and S.M. Block. 1999. Single kinesin molecules studied with a molecular force clamp. *Nature.* 400:184–189. doi:10.1038/22146.

Wang, W., L. Cao, C. Wang, B. Gigant, and M. Knossow. 2015. Kinesin, 30 years later: recent insights from structural studies. *Protein Sci.* 24:1047–1056. doi:10.1002/pro.2697.

Whitmarsh, A.J. 2006. The JIP family of MAPK scaffold proteins. *Biochem. Soc. Trans.* 34:828–832. doi:10.1042/BST0340828.

Wickstead, B., K. Gull, and T.A. Richards. 2010. Patterns of kinesin evolution reveal a complex ancestral eukaryote with a multifunctional cytoskeleton. *BMC Evol. Biol.* 10:110. doi:10.1186/1471-2148-10-110.

Wilson, M.H., and E.L.F. Holzbaur. 2015. Nesprins anchor kinesin-1 motors to the nucleus to drive nuclear distribution in muscle cells. *Development.* 142:218–228. doi:10.1242/dev.114769.

Woźniak, M.J., and V.J. Allan. 2006. Cargo selection by specific kinesin light chain 1 isoforms. *EMBO J.* 25:5457–5468. doi:10.1038/sj.emboj.7601427.

Xia, C.-H., E.A. Roberts, L.-S. Her, X. Liu, D.S. Williams, D.W. Cleveland, and L.S.B. Goldstein. 2003. Abnormal neurofilament transport caused by targeted disruption of neuronal kinesin heavy chain KIF5A. *J. Cell Biol.* 161:55–66. doi:10.1083/jcb.200301026.

Xu, M., Y. Gu, J. Barry, and C. Gu. 2010. Kinesin I transports tetramerized Kv3 channels through the axon initial segment via direct binding. *J. Neurosci.* 30:15987–16001. doi:10.1523/JNEUROSCI.3565-10.2010.

Yildiz, A., M. Tomishige, R.D. Vale, and P.R. Selvin. 2004. Kinesin walks hand-over-hand. *Science.* 303:676–678. doi:10.1126/science.1093753.

Yip, Y.Y., S. Pernigo, A. Sanger, M. Xu, M. Parsons, R.A. Steiner, and M.P. Dodding. 2016. The light chains of kinesin-1 are autoinhibited. *Proc. Natl. Acad. Sci. U.S.A.* 113:2418–2423. doi:10.1073/pnas.1520817113.

3 The Kinesin-2 Family
Transporters

William O. Hancock

CONTENTS

The Kinesin-2 family carry out transport functions and members of this family are crucial for intraflagellar transport. The family can be divided into the homodimeric and heterotrimeric members.

3.1   EXAMPLE FAMILY MEMBERS

Homodimeric

Mammalian: KIF17
Caenorhabditis elegans: Osm3

Heterotrimeric

Mammalian: KIF3A/B
Drosophila melanogaster: KLP64D/68D
C. elegans: KLP11/20
Chlamydomonas: FLA8/10

3.2   STRUCTURAL INFORMATION

The Kinesin-2 heavy chain (HC) contains an N-terminal motor domain, a central coiled-coil domain and a C-terminal tail domain. Similar to Kinesin-1, the central domain contains a proximal "neck-coil" dimerisation region, and then a long coiled-coil region broken by a hinge domain that allows the motor to fold up and the tail to inhibit the motor domains in the absence of cargo (Hammond et al. 2010).

The Kinesin-2 family can be divided into two classes, the homodimeric members and the heterotrimeric members (Figure 3.1; Table 3.1) (Scholey 2013). Homodimeric Kinesin-2s contain two identical HCs that homodimerise, but currently no accessory light chains are known. Heterotrimeric Kinesin-2s contain two HCs that heterodimerise through a coiled-coil, and a single KAP3 domain that binds to the tail of the motor.

The two HCs in heterotrimeric Kinesin-2 contain a region of opposing charges in the proximal coiled-coil region, that had been hypothesised to drive heterodimerisation, but it was subsequently shown that heterodimerisation is instead driven by a trigger sequence at the C-terminal end of the coiled-coil domain (De Marco et al. 2001). In heterotrimeric Kinesin-2, the KAP3 subunit binds to the globular C-terminal tail of the HCs.

One notable structural detail is that the neck linker of Kinesin-2 that connects the catalytic core to the dimerisation domain, and which undergoes important structural changes during stepping, is three residues longer than the corresponding Kinesin-1 neck linker (17 rather than 14 residues). This extension was shown to explain the shorter unloaded run length of Kinesin-1 relative to Kinesin-2 (Shastry and Hancock 2010).

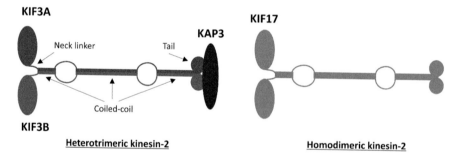

FIGURE 3.1 Diagram of Kinesin-2 structure. The family can be separated into heterotrimeric and homodimeric members, each of which has two N-terminal motor domains, a coiled-coil domain, and a C-terminal tail domain.

TABLE 3.1
Kinesin-2 Nomenclature Across Species

| Species | Heterotrimeric subunits | | | Homodimeric |
	α	β	κ	γ
Vertebrate	KIF3A	KIF3B, KIF3C	KAP3/KIFAP3	KIF17
Sea urchin	KRP85	KRP95	KAP	
Drosophila	KLP64D	KLP68D	KAP	
C. elegans	KLP20	KLP11	KAP1	OSM-3
Chlamydomonas	FLA10	FLA8	FLA3	

3.3 FUNCTIONAL PROPERTIES

The Kinesin-1, 2 and 3 families are the core transport motors in the kinesin superfamily. As such, Kinesin-2 motors are generally processive, but they are generally slower than Kinesin-1 and 3 (although there are exceptions), and the mechanochemical paradigms set out for Kinesin-1 generally fit well for Kinesin-2 motors. Although they carry out a range of transport functions, two notable Kinesin-2 functions are vesicle transport in neurons and intraflagellar transport in flagella and cilia (Figure 3.2). One notable property of heterotrimeric Kinesin-2 is that its detachment rate is very sensitive to load. Andreasson et al. (2015) compared Kinesin-1 and Kinesin-2 and found that, whereas Kinesin-1 would step against substantial loads up to its stall force, the Kinesin-2, KIF3A/B, detached readily, even at low loads (Andreasson et al. 2015). Other studies supported this claim (Arpag et al. 2014, Milic et al. 2017, Schroeder et al. 2012). Interestingly, the homodimeric Kinesin-2, KIF17, was shown to be much more resistant to loads, and to have load-dependent detachment properties very similar to Kinesin-1 (Milic et al. 2017). This propensity to detach under load seems surprising for a transport motor that competes against dynein; however, it was shown that the rate of motor reattachment (and initial attachment from solution) is of the order of four times faster for Kinesin-1 than Kinesin-2 (Feng et al. 2018). Thus, it can be argued that heterotrimeric Kinesin-2 functions primarily as a tether that keeps vesicles close to the microtubule, while Kinesin-1s attached to the same vesicle are responsible for the bulk of the transport and force generation.

One interesting characteristic of some Kinesin-2 family members is their propensity to spiral around microtubules with a left-handed pitch (Brunnbauer et al. 2012). This property, which was measured using beads coated with multiple motors, was more pronounced in less processive motors, though a subset of processive motors also showed spiralling, and spiralling did not require two dissimilar heads (Figure 3.3). Structurally, the neck linker domain was the crucial domain that controlled the spiralling behaviour. Introducing a long and flexible Gly-Ser insert into the neck linker of a Kinesin-1 conferred spiralling behaviour, whereas introducing a cysteine crosslink into a Kinesin-2 abolished the spiralling behaviour.

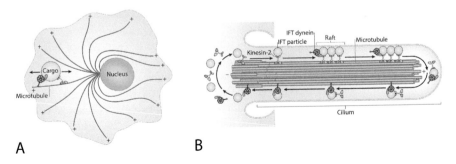

FIGURE 3.2 (A) Kinesin-2 transports cargo toward the cell periphery along cytoplasmic microtubules and (B) carries cargo towards the tips of cilia and flagella. Taken from Hancock (2014).

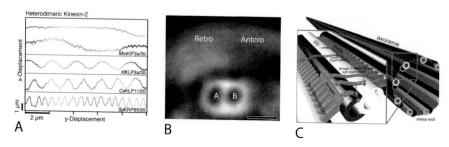

FIGURE 3.3 Kinesin-2 behaviour in intraflagellar transport that limits head-on traffic jams. (A) Heterotrimeric Kinesin-2s were shown to take spiral tracks around microtubules, with less processive motors having tighter spiral pitches. Taken from Brunnbauer et al. (2012). (B) Cryogenic electron microscopy (Cryo-EM) image of one flagellar microtubule doublet, with superimposed position of retrograde-transported cargo, driven by dynein along A-tubules, and anterograde-transported cargo, driven by Kinesin-2 along B-tubules. Scale bar 25 nm. Taken from Stepanek and Pigino (2016). (C) Model of how spiralling behaviour of Kinesin-2 brings anterograde-transported cargo away from the A-tubule. Inter-doublet connections on the B-tubule limit further spiralling and keep the cargo away from the A-tubule, thus preventing collisions with cargo moving retrograde along the A-tubule. Taken from Stepp et al. (2017).

3.4 PHYSIOLOGICAL ROLE

The paradigmatic physiological role of the Kinesin-2 family is anterograde intraflagellar transport (IFT). However, heterotrimeric and homodimeric Kinesin-2s have many other cytoplasmic transport functions in cells. In virtually every ciliar and flagellar structure known, heterotrimeric Kinesin-2s are required for proper development and maintenance of the cilia or flagella. One hallmark of IFT is that cargoes move continuously to the tip of the cilium, and, after a waiting time, move continuously back to the cell body. This continuous unidirectional transport, with a single directional switch, differs from axonal and most other cytoplasmic transport, where cargoes pause, switch directions and periodically diffuse freely before engaging with another microtubule (Hancock 2014). This uninterrupted movement likely results from the fact that anterograde transport occurs on the B-tubule of the axonemal microtubule doublet, whereas retrograde transport occurs on the A-tubule (Figure 3.3) (Stepanek and Pigino 2016).

An interesting model system that highlights the physiological role of Kinesin-2 transport, is the sensory cilium in *C. elegans* nematodes. This structure differs from most cilia and flagella because, in addition to the doublet axonemal microtubules that are found in the central region of the cilium, the B-tubules do not extend to the tip of the cilium, leaving a distal segment containing only singlet A-tubules. In these sensory cilia, the IFT rafts contain both heterodimeric KLP11/20 and homodimeric OSM3 Kinesin-2 motors, and the trajectories of IFT particles include a slower velocity region in the middle of the cilium and a faster velocity region in the last few microns at the end of the cilium (Snow et al. 2004). This behaviour was traced to the two motors; along the proximal microtubule doublets, the slower KLP11/20 and the faster OSM3 worked together, whereas only the faster OSM3 moved along the distal singlet microtubules (Snow et al. 2004). This behaviour presumably results from the fact that the heterotrimers move along the B-tubule in doublets (Stepanek and Pigino 2016), whereas the homodimers either preferentially walk on A-tubules or walk on both A- and B-tubules.

In axons, heterotrimeric Kinesin-2s carry fodrin-bound vesicles and choline acetyltransferase receptors out to synapses, whereas, in dendrites, homodimeric Kinesin-2s transport N-methyl-D-aspartate (NMDA) receptors and voltage-gated potassium channels (Scholey 2013). In rod photoreceptor cells, heterotrimeric Kinesin-2 transports opsin to the outer segment, and homodimeric Kinesin-2 has been implicated in ciliogenesis in vertebrate photoreceptor cells (Scholey 2013). In non-neuronal cells, Kinesin-2 transports a range of vesicles, including Golgi-derived vesicles, melanosomes, late endosomes and lysosomes (Scholey 2013). Kinesin-2s are also thought to transport HIV viruses in macrophages (Gaudin et al. 2012) and mRNA in *Xenopus* oocytes (Messitt et al. 2008).

Kinesin-2 motors are also implicated in establishing and maintaining cell polarity. In dendrites of *Drosophila* dopamine neurons, a complex of Kinesin-2, EB1 and adenomatous polyposis coli (APC) was found to be necessary for maintaining uniform minus-end-out microtubule organisation (Mattie et al. 2010). It was proposed that EB1 tracks the plus-ends of microtubules growing into branch points, and that Kinesin-2 generates force along the existing microtubules to guide the growing plus-ends toward the cell body, thus maintaining the minus-end-out orientation. In support of this mechanism, *in vitro* reconstructions showed that the strength of interactions between microtubules and both EB1 and Kinesin-2 was sufficient to bend growing microtubules (Chen, Rolls, and Hancock 2014). Kinesin-2s have also been shown to associate with beta-catenin and cadherins, implicating these motors in cell adhesion and cell polarity (Jimbo et al. 2002, Murawala et al. 2009).

3.5 INVOLVEMENT IN DISEASE

Based on the diverse cellular functions of Kinesin-2, it is not surprising that defects in Kinesin-2-based transport are implicated in a diverse range of diseases. The primary cilium, which is found on most cells in the human body, is a sensory organelle that carries out a range of signalling functions, including those in the Hedgehog and Wnt pathways. The importance of intraflagellar transport for building and maintaining these cilia cannot be overstated. One of the most striking examples of the importance of Kinesin-2-driven transport is the observation that mouse knockouts of the gene encoding the KIF3B subunit of heterotrimeric Kinesin-2 displayed *sinus invertus* – randomisation of the left-right body axis asymmetry – as well as heart and other developmental defects (Nonaka et al. 1998). This phenotype was traced to a structure called the nodal cilium, that rotates and generates directional fluid flow in the developing embryo. In KIF3B mutants, IFT defects led to immotile cilia and resulted in a lack of directional flow of morphogens that specify asymmetrical left-right development of the embryo.

In addition to ciliary defects, Kinesin-2 has been linked to cell migration in breast cancer cells (Lukong and Richard 2008), and a truncation of homodimeric Kinesin-2 KIF17 has been linked to schizophrenia (Scholey 2013, Tarabeux et al. 2010).

REFERENCES

Andreasson, J. O., S. Shastry, W. O. Hancock, and S. M. Block. 2015. "The mechanochemical cycle of mammalian kinesin-2 KIF3A/B under load." *Curr Biol* 25(9):1166–75. doi:10.1016/j.cub.2015.03.013.

Arpag, G., S. Shastry, W. O. Hancock, and E. Tuzel. 2014. "Transport by populations of fast and slow kinesins uncovers novel family-dependent motor characteristics important for in vivo function." *Biophys J* 107(8):1896–904. doi:10.1016/j.bpj.2014.09.009.

Brunnbauer, M., R. Dombi, T. H. Ho, M. Schliwa, M. Rief, and Z. Okten. 2012. "Torque generation of kinesin motors is governed by the stability of the neck domain." *Mol Cell* 46(2):147–58. doi:10.1016/j.molcel.2012.04.005.

Chen, Y., M. M. Rolls, and W. O. Hancock. 2014. "An EB1-kinesin complex is sufficient to steer microtubule growth in vitro." *Curr Biol* 24(3):316–21. doi:10.1016/j.cub.2013.11.024.

De Marco, V., P. Burkhard, N. Le Bot, I. Vernos, and A. Hoenger. 2001. "Analysis of heterodimer formation by Xklp3A/B, a newly cloned kinesin-II from Xenopus laevis." *EMBO J* 20(13):3370–9.

Feng, Q., K. J. Mickolajczyk, G. Y. Chen, and W. O. Hancock. 2018. "Motor reattachment kinetics play a dominant role in multimotor-driven cargo transport." *Biophys J* 114(2):400–9. doi:10.1016/j.bpj.2017.11.016.

Gaudin, R., B. C. de Alencar, M. Jouve, S. Berre, E. Le Bouder, M. Schindler, A. Varthaman, F. X. Gobert, and P. Benaroch. 2012. "Critical role for the kinesin KIF3A in the HIV life cycle in primary human macrophages." *J Cell Biol* 199(3):467–79. doi:10.1083/jcb.201201144.

Hammond, J. W., T. L. Blasius, V. Soppina, D. Cai, and K. J. Verhey. 2010. "Autoinhibition of the kinesin-2 motor KIF17 via dual intramolecular mechanisms." *J Cell Biol* 189(6):1013–25. doi:jcb.201001057 [pii]10.1083/jcb.201001057.

Hancock, W. O. 2014. "Bidirectional cargo transport: moving beyond tug of war." *Nat Rev Mol Cell Biol* 15(9):615–28. doi:10.1038/nrm3853.

Jimbo, T., Y. Kawasaki, R. Koyama, R. Sato, S. Takada, K. Haraguchi, and T. Akiyama. 2002. "Identification of a link between the tumour suppressor APC and the kinesin superfamily." *Nat Cell Biol* 4(4):323–7. doi:10.1038/ncb779ncb779 [pii].

Lukong, K. E., and S. Richard. 2008. "Breast tumor kinase BRK requires kinesin-2 subunit KAP3A in modulation of cell migration." *Cell Signal* 20(2):432–42. doi:10.1016/j.cellsig.2007.11.003.

Mattie, F. J., M. M. Stackpole, M. C. Stone, J. R. Clippard, D. A. Rudnick, Y. Qiu, J. Tao, D. L. Allender, M. Parmar, and M. M. Rolls. 2010. "Directed microtubule growth, +TIPs, and kinesin-2 are required for uniform microtubule polarity in dendrites." *Curr Biol* 20(24):2169–77. doi:S0960-9822(10)01516-2 [pii]10.1016/j.cub.2010.11.050.

Messitt, T. J., J. A. Gagnon, J. A. Kreiling, C. A. Pratt, Y. J. Yoon, and K. L. Mowry. 2008. "Multiple kinesin motors coordinate cytoplasmic RNA transport on a subpopulation of microtubules in Xenopus oocytes." *Dev Cell* 15(3):426–36. doi:10.1016/j.devcel.2008.06.014.

Milic, B., J. O. L. Andreasson, D. W. Hogan, and S. M. Block. 2017. "Intraflagellar transport velocity is governed by the number of active KIF17 and KIF3AB motors and their motility properties under load." *Proc Natl Acad Sci U S A* 114(33):E6830–8. doi:10.1073/pnas.1708157114.

Murawala, P., M. M. Tripathi, P. Vyas, A. Salunke, and J. Joseph. 2009. "Nup358 interacts with APC and plays a role in cell polarization." *J Cell Sci* 122(Pt 17):3113–22. doi:10.1242/jcs.037523.

Nonaka, S., Y. Tanaka, Y. Okada, S. Takeda, A. Harada, Y. Kanai, M. Kido, and N. Hirokawa. 1998. "Randomization of left-right asymmetry due to loss of nodal cilia generating leftward flow of extraembryonic fluid in mice lacking KIF3B motor protein." *Cell* 95(6):829–37.

Scholey, J. M. 2013. "Kinesin-2: a family of heterotrimeric and homodimeric motors with diverse intracellular transport functions." *Annu Rev Cell Dev Biol* 29:443–69. doi:10.1146/annurev-cellbio-101512-122335.

Schroeder, H. W. , 3rd, A. G. Hendricks, K. Ikeda, H. Shuman, V. Rodionov, M. Ikebe, Y. E. Goldman, and E. L. Holzbaur. 2012. "Force-dependent detachment of kinesin-2 biases track switching at cytoskeletal filament intersections." *Biophys J* 103(1):48–58. doi:10.1016/j.bpj.2012.05.037.

Shastry, S., and W. O. Hancock. 2010. "Neck linker length determines the degree of processivity in Kinesin-1 and Kinesin-2 motors." *Curr Biol* 20:939–43.

Snow, J. J., G. Ou, A. L. Gunnarson, M. R. Walker, H. M. Zhou, I. Brust-Mascher, and J. M. Scholey. 2004. "Two anterograde intraflagellar transport motors cooperate to build sensory cilia on C. elegans neurons." *Nat Cell Biol* 6(11):1109–13.

Stepanek, L., and G. Pigino. 2016. "Microtubule doublets are double-track railways for intraflagellar transport trains." *Science* 352(6286):721–4. doi:10.1126/science.aaf4594.

Stepp, W. L., G. Merck, F. Mueller-Planitz, and Z. Okten. 2017. "Kinesin-2 motors adapt their stepping behavior for processive transport on axonemes and microtubules." *EMBO Rep* 18(11):1947–56. doi:10.15252/embr.201744097.

Tarabeux, J., N. Champagne, E. Brustein, F. F. Hamdan, J. Gauthier, M. Lapointe, C. Maios, A. Piton, D. Spiegelman, E. Henrion, Team Synapse to Disease, B. Millet, J. L. Rapoport, L. E. Delisi, R. Joober, F. Fathalli, E. Fombonne, L. Mottron, N. Forget-Dubois, M. Boivin, J. L. Michaud, R. G. Lafreniere, P. Drapeau, M. O. Krebs, and G. A. Rouleau. 2010. "De novo truncating mutation in Kinesin 17 associated with schizophrenia." *Biol Psychiatry* 68(7):649–56. doi:10.1016/j.biopsych.2010.04.018.

4 The Kinesin-3 Family
Long-Distance Transporters

Nida Siddiqui and Anne Straube

CONTENTS

The Kinesin-3s are a family of cargo transporters. They typically display highly processive plus-end-directed motion, either as dimers or in teams, formed via interaction with cargo.

4.1 EXAMPLE FAMILY MEMBERS

Mammalian: KIF1A, KIF1B, KIF1C, KIF13A, KIF13B, KIF14, KIF16B
Drosophila melanogaster: UNC-104
Caenorhabditis elegans: UNC-104, KLP6

4.2 STRUCTURAL INFORMATION

The Kinesin-3 family is classified into five subfamilies, named Kinesin-3A/KIF28, -3B/KIF16, -3C/KIF1, -3D/KIF13 and -3E/KIF14 (Miki et al., 2005, Wickstead and Gull, 2006). Members of the family possess an N-terminal motor domain, followed by a neck domain, a subfamily-specific forkhead-associated (FHA) domain (Westerholm-Parvinen et al., 2000), typically several regions of predicted coiled-coil and a diverse C-terminal tail containing short coiled coils, with lipid- and protein-interaction regions that aid in cargo- as well as adapter-binding (Figure 4.1A).

The Kinesin-3 motor domain contains a characteristic stretch of lysine residues in Loop 12, known as the K-loop. This loop forms part of the microtubule-binding surface and interacts with the glutamate-rich C-terminal tail of β-tubulin. This

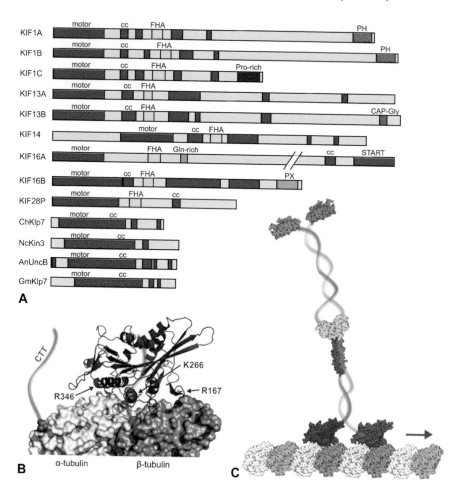

FIGURE 4.1 Structure of members of the Kinesin-3 family. (A) Domain organisation in Kinesin-3s shows the characteristic N-terminal location of the motor domain, FHA domain and tail, with several short coiled-coil (CC) regions in addition to a variety of protein- or lipid-interaction motifs. (B) KIF1A motor domain (dark red) bound to tubulin (grey) with C-terminal tubulin tails indicated in green. Key residues specific for the Kinesin-3 family that increase processivity of KIF1A (Atherton et al., 2014, Scarabelli et al., 2015) are highlighted in blue. PDB: 4UXP. (C) Composite of a dimeric Kinesin-3 with motor domains (red) from KIF1A, FHA domain (yellow) and CC1 (blue) from KIF13B and PX domain (orange) from KIF16B. PDB: 4UXP, 5DJO and 2V14, using Illustrate (Goodsell et al., 2019).

interaction has been reported to influence the microtubule on-rate for the 3B, 3C and 3D subfamilies (Soppina and Verhey, 2014, Rogers et al., 2001, Lessard et al., 2019, Matsushita et al., 2009). The K-loop has also been proposed to enable diffusive movement on the microtubules for a subset of Kinesin-3s (Okada and Hirokawa, 1999, 2000), as well as facilitating Kinesin-3 motors to work in teams (Rogers et al., 2001, Soppina and Verhey, 2014). Comparison of high-resolution cryo electron microscopy structures of Kinesin-1 (KIF5A) and Kinesin-3 (KIF1A) motor domains bound to microtubules in different nucleotide states, combined with molecular dynamics simulations, suggests that multiple amino acid differences spread over the microtubule-binding interface contribute to the 200-fold greater affinity of Kinesin-3 for microtubules compared with Kinesin-1 (Scarabelli et al., 2015, Atherton et al., 2014). This greater affinity results in increased processivity of dimeric Kinesin-3 motors (Atherton et al., 2014). The key processivity-determining residues are Arg167 in loop 8, Lys266 in loop 11 and Arg346 in α-helix 6 of KIF1A (Scarabelli et al., 2015) (Figure 4.1B).

Most Kinesin-3s act as dimers, with coiled-coil regions enabling motor dimerisation as well as interaction with binding partners (Peckham, 2011) (Figure 4.1C). The neck coil has been reported to drive dimerisation for KIF1A, KIF13A and KIF13B (Hammond et al., 2009, Soppina et al., 2014), whereas for KIF1C, the fourth coiled-coil domain is sufficient to promote dimerisation (Dorner et al., 1999). For the *C. elegans* Kinesin-3, Unc-104, coiled-coil regions mediate interaction with dynein/dynactin subunits (Chen et al., 2019). Recent X-ray crystallographic structures of monomeric and dimeric conformations of KIF13B confirm that coiled-coil 1 is crucial for maintaining autoinhibition of the motor domain (Ren et al., 2018).

FHA domains are phospho-peptide recognition domains, found in several regulatory proteins, that mediate protein–protein interactions. In Kinesin-3s, the FHA domain provides structural support, as well as mediating cargo interactions. Cargo binding by KIF13B is regulated by the phosphorylation of T506 in the FHA domain by the cyclin-dependent kinase 5 (Cdk-5), which allows binding of transient receptor potential vanilloid 1 (TRPV1) (Xing et al., 2012). A change in the susceptibility of mice to anthrax lethal toxin was reported as a result of a point mutation in the FHA domain of KIF1C, that is likely to alter the folding, which further highlights the functional importance of the domain (Durocher and Jackson, 2002, Watters et al., 2001).

The C-terminal region is diverse, but several Kinesin-3s contain a lipid-interaction domain (Figure 4.1A). KIF1A and KIF1B have a pleckstrin homology (PH) domain that is important for binding cargo vesicles (Xue et al., 2010). KIF16A contains a StAR-related lipid transfer (START) lipid/sterol-binding domain (Torres et al., 2011), whereas KIF16B contains a phosphoinositide-binding structural domain (PX), which is involved in the trafficking of early endosomes (Blatner et al., 2007, Hoepfner et al., 2005). KIF1C has a proline-rich region in the C-terminal region, which interacts with the cargo adapter protein BICDR1, 14-3-3 proteins and Rab6 (Schlager et al., 2010, Dorner et al., 1999, Lee et al., 2015b). The C-terminal region of KIF13B contains a CAP-Gly domain, and mice expressing a truncated KIF13B, lacking the CAP-Gly domain, exhibit reduced uptake of LRP-1 (LDL receptor-related protein-1) (Mills et al., 2019).

4.3 FUNCTIONAL PROPERTIES

Kinesin-3s are a family of transporters and most of their members are implicated in long-distance cargo transport. In general, dimeric Kinesin-3 motors are highly processive (Table 4.1). This is achieved by the presence of the K-loop and other structural features of the motor domain, which maintain a stable interaction between

TABLE 4.1

Reported Speed and Run Length of Kinesin-3 Motors

	Average speed	Run length	Nature of construct/assay	References
KIF1	1.4 μm/s	3.4 μm	DCVs in neurons	Lipka et al. (2016)
KIF1A	1.2 μm/s		Full-length purified motor, gliding assay	Okada et al. (1995)
	0.65 μm/s	0.44 μm	Full-length motor, SMMA using COS-7 lysates	Hammond et al. (2009)
	1.5 μm/s	2.6 μm	Truncated motor, SMMA using COS-7 lysates	Soppina et al. (2014)
	2.45 μm/s	9.8 μm	Truncated motor, LZ, SMMA using COS-7 lysates	Soppina et al. (2014)
	0.08 μm/s		Truncated motor in COS-7 cells	Lipka et al. (2016)
Unc104	1.6 μm/s	1.5 μm	Truncated motor, LZ, SMMA	Tomishige et al. (2002)
KIF1B	0.17 μm/s		Truncated motor in COS-7 cells	Lipka et al. (2016)
KIF1C	0.28 μm/s		Truncated motor in COS-7 cells	Lipka et al. (2016)
	2.0 μm/s		Truncated motor, gliding assay	Rogers et al. (2001)
	0.45 μm/s	8.6 μm	Full-length purified motor, SMMA	Siddiqui et al. (2019)
	0.73 μm/s	15.8 μm	Full-length purified motor, SMMA	Kendrick et al. (2019)
KIF13A	0.1–0.3 μm/s		Full-length purified motor, gliding assay	Nakagawa et al. (2000)
	1.4 μm/s	10 μm	Truncated motor, SMMA using COS-7 lysates	Soppina et al. (2014)
KIF13B	0.08 μm/s		Truncated motor in COS-7 cells	Lipka et al. (2016)
	1.3 μm/s	10 μm	Truncated motor, SMMA using COS-7 lysates	Soppina et al. (2014)
Khc73	1.5 μm/s	1 μm	Truncated motor SMMA	Huckaba et al. (2011)
KIF16B	0.97 μm/s	9.5 μm	Truncated motor, SMMA using COS-7 lysates	Soppina et al. (2014)
	0.09 μm/s		Truncated motor in COS-7 cells	Lipka et al. (2016)

Note: SMMA, single-molecule motility assay; LZ, dimerised using leucine zipper; DCVs, dense core vesicles labelled with neuropeptide Y; COS-7, immortalised African Green Monkey cells.

the motor domain and the microtubule throughout the ATPase cycle (Atherton et al., 2014, Okada and Hirokawa, 2000, Scarabelli et al., 2015). Full-length KIF1A is monomeric and can generate about 0.15 pN force (Okada et al., 2003) and very slow (0.15 μm/s) plus-end-directed movement along microtubules (Okada and Hirokawa, 1999). When dimerised or acting in teams, Kinesin-3 motors are 100 times faster and stronger (Okada et al., 2003, 1995, Oriola and Casademunt, 2013). Indeed, full-length KIF1A, dimerised by addition of a leucine zipper, moves with an average speed of 1.3 µm/s and an average run length of 6 µm (Lessard et al., 2019), and dimerised Unc-104 can generate forces of up to 6 pN (Tomishige et al., 2002). Full-length KIF1C forms a natural dimer and has an average speed of 0.5–0.7 µm/s and an average run length of 9–16 µm (Siddiqui et al., 2019, Kendrick et al., 2019). Removal of its autoinhibitory stalk domain results in a hyperactive KIF1C, with an average speed of 1.2 µm/s (Siddiqui et al., 2019), similar to the transport speed of KIF1C-dependent cargo in cells.

4.3.1 Autoinhibition of Kinesin-3 Motors and Their Activation

Two mechanisms for Kinesin-3 motor regulation have been described. Most Kinesin-3s undergo a monomer-to-dimer transition upon cargo binding, but some are autoinhibited dimers, in which the stalk blocks motor activity until a cargo binds and releases the inhibition (Siddiqui and Straube, 2017).

Regulation by monomer-to-dimer switch is mediated by intramolecular interactions between the neck and tail regions, that hold the kinesin in a monomeric, inactive state. Upon cargo binding, the intramolecular interaction between the neck and tail regions is disrupted and these motors dimerise (Soppina et al., 2014, Tomishige et al., 2002, Okada and Hirokawa, 1999). This can be observed for Unc-104 and KIF1A, which are largely inactive in single motility assays, with intermolecular interactions of the neck coil segment regulating the monomer-dimer transition (Hammond et al., 2009, Al-Bassam et al., 2003). Mutations in the first coiled-coil segment of KIF1A also result in activation of the motor in cells (Yue et al., 2013, Huo et al., 2012). In the KIF13 subfamily, neck coil 1 and coiled-coil 1 are important for the regulation of dimerisation, and deletion of a proline residue at the junction of these domains results in processive dimeric motors (Ren et al., 2016, Soppina et al., 2014).

Autoinhibited dimer regulation occurs for Kinesin-3s that are stable dimers. The stalk, usually a region in the middle of the molecule, interacts with the motor domain to form an autoinhibited state. When an adapter protein or cargo binds to the stalk region, the motor is released from this autoinhibited state. For KIF1C, the binding of PTPN21 or Hook3 to the stalk releases the motor domain and activates intracellular transport (Siddiqui et al., 2019). A similar mechanism seems to regulate KIF13B, which is autoinhibited in solution, but active in a gliding assay. This is likely because binding of the C-terminal tail to the surface resembles the cargo-bound state. In cells, KIF13B is activated when its cargo – human discs large (hDlg) tumour suppressor – binds to the stalk domain and relieves inhibition (Yamada et al., 2007). KIF13B is also regulated by phosphorylation at S1381 and S1410 by Par1b/MARK2 (microtubule affinity-regulating kinase). This allows 14-3-3β binding to the stalk and promotes the intramolecular interaction of KIF13B motor and stalk domains. Consequently, KIF13B microtubule binding is impaired, resulting in the dispersal of the motor in the cytoplasm and a reduction in cell protrusion and axon formation (Yoshimura et al., 2010).

KIF16B exhibits a mechanism of autoinhibition whereby the monomeric motor is held in an autoinhibited conformation by intramolecular interactions of the second and third coiled-coil with the motor domain (Farkhondeh et al., 2015). Using Förster resonance energy transfer (FRET), it was observed that these motors dimerise on the surface of endosomes (Soppina et al., 2014), so that cargo binding both releases the autoinhibition and facilitates dimerisation of the motor.

4.4 PHYSIOLOGICAL ROLES

Cargoes have been identified for most Kinesin-3 family members (Table 4.2, online material). These range from organelles, such as mitochondria, lysosomes and endosomes, to specific proteins, mRNAs and viral particles. Due to their speed and processivity, Kinesin-3 motors are implicated primarily in neuronal transport.

The movement of dense core vesicles in neurons depends on KIF1A and KIF1C and occurs at an average speed of 1.4 µm/s (Lipka et al., 2016). Integrin-containing vesicles are KIF1C-dependent cargoes that reach top speeds of 2 µm/s and move at an average rate of 0.4 µm/s in human retinal pigment epithelial cells (Theisen et al., 2012). In COS-7 cells, using a rapamycin analogue (rapalogue)-inducible peroxisome-trafficking assay, six out of eight kinesin-3 members were found to significantly enhance transport.

In C. elegans, the KIF1A orthologue Unc-104 is a neuron-specific fast anterograde transporter of synaptic vesicle precursors (Okada et al., 1995). Unc-104 cargoes include synaptotagmin, synaptophysin and synaptobrevin-1 (Nonet, 1999, Okada et al., 1995).

Other than neurons, KIF1C has been suggested to mediate Golgi-to-endoplasmic reticulum transport (Dorner et al., 1998) and to maintain Golgi structure (Lee et al., 2015b). In migrating cells, KIF1C transports integrins and is responsible for the maintenance of cell tails and the maturation of focal adhesion sites (Theisen et al., 2012). In macrophages and vascular smooth muscle cells, KIF1C contributes to the formation and regulation of actin-rich podosome structures (Efimova et al., 2014, Kopp et al., 2006). Members of the Kinesin-3 family also play a role in cell division. KIF16A enables the formation of a bipolar mitotic spindle by tethering the pericentriolar material (PCM) to the daughter centriole during mitosis, thereby preventing PCM fragmentation (Torres et al., 2011). KIF13A plays a central role in cytokinesis by translocating FYVE-CENT, a component of the cell abscission machinery, to the spindle midzone (Sagona et al., 2010). KIF14 is upregulated during mitosis and localised to the spindle midzone, with its depletion causing cytokinesis failure and cell death (Carleton et al., 2006). In Ustilago maydis, deletion of the sole kinesin-3 leads to a cell separation defect (Wedlich-Soldner, 2002).

The localisation of vesicles is largely controlled by members of the Rab family of GTPases (Zerial and McBride, 2001), and Kinesin-3 members interact with many different Rabs. KIF1A and KIF1Bβ transport Rab3, a synaptic vesicle protein that controls exocytosis of synaptic vesicles along the axon. KIF1C interacts with Rab6 at two sites. Rab6 binding to the motor domain disrupts the motor–microtubule interaction (Lee et al., 2015b), whereas binding to the C-terminus is proposed to mediate cargo loading and subsequent activation. KIF1C also transports Rab11-positive vesicles for the recycling of integrins (Theisen et al., 2012), but the role of Rab11 in controlling the activity of KIF1C remains to be understood. Another motor protein that binds Rab11-positive vesicles is KIF13A, that controls endosomal sorting and recycling of Rab11-positive endosomal cargo (Delevoye et al., 2014). KIF13A and KIF13B were also identified as interacting partners for Rab10 and the Rab10–KIF13 complex was implicated in the formation of tubular endosomes (Etoh and Fukuda, 2019). This suggests that the KIF13 motors might be able to interact with several Rabs to mediate consecutive steps of cargo sorting, endosome biogenesis and transport.

In non-neuronal cells, KIF16B transports Rab5-positive early endosomes and Rab14-positive vesicles (Hoepfner et al., 2005, Ueno et al., 2011).

An emerging field in Kinesin-3 biology is their cooperation with dynein in bidirectional cargo transport. KIF1C is activated by the cargo adapter Hook3 (Siddiqui et al., 2019) and also interacts with BICDR-1 via its proline-rich tail region (Schlager et al., 2010). Both Hook3 and BICDR-1 are activators of the minus-end-directed motor dynein, and Hook3 can bind simultaneously to KIF1C and dynein (Redwine et al., 2017, Urnavicius et al., 2018, Kendrick et al., 2019). The possibility of complexes containing opposite-polarity motors opens interesting new possibilities in the regulation of Kinesin-3 activity.

It is becoming clear that no typical mechanism for regulating Kinesin-3 motor activity exists and each motor–cargo combination needs to be studied separately to understand how loading of a specific cargo modulates motor activity and thereby determines its cellular distribution.

4.4.1 Preference for Subsets of Microtubule Tracks

Tubulin undergoes a diverse range of post-translational modifications, usually after it polymerises into microtubules. These modifications occur predominantly on the C-terminal tails of both α- and β-tubulin (Magiera and Janke, 2014). The affinity of motors for microtubules can be altered by these modifications, ultimately acting as guides for motor transport (Janke, 2014). The Kinesin-3 family-specific K-loop is proposed to interact with the C-terminal tail (CTT) of β-tubulin, so it is expected that changes in this region would impact Kinesin-3 binding. In ROSA22 mice, knockdown of the polyglutamylase PGs1 resulted in reduced localisation of KIF1A to neurites (Ikegami et al., 2007). Single-molecule imaging of dimerised full-length KIF1A implicates the K-loop in engaging with polyglutamylated CTTs during pause events, thereby linking several processive runs and resulting in the super-processive behaviour typical of Kinesin-3s (Lessard et al., 2019). However, some reports suggest that KIF1A and KIF1Bβ drive lysosomal transport preferentially along tyrosinated (i.e. non-modified) microtubules (Guardia et al., 2016) and that the tubulin deglutamylase CCPP-1 positively regulates the ciliary localisation of *C. elegans* KLP-6 (O'Hagan et al., 2011). In primary human macrophages, the peripheral localisation of KIF1C is negatively regulated by acetylation (Bhuwania et al., 2014), suggesting that acetylation reduces motor activity. For the fungal Kinesin-3 UncA, the tail is necessary and sufficient to guide the motor to selectively recognise detyrosinated microtubules (Zekert and Fischer, 2009; Seidel et al., 2012). These data suggest that most Kinesin-3s recognise post-translational tubulin modifications.

Kinesin-3 interaction with microtubules can also be regulated via the presence of other microtubule-associated proteins (MAPs). MAP7 and Tau have been shown to negatively regulate KIF1A activity, observed as a reduction in the landing rate in the presence of MAP7 or Tau proteins (Monroy et al., 2018). MAP9 was shown to promote KIF1A motility by interacting with the K-loop of KIF1A, mediated by a transient ionic interaction (Monroy et al., 2019). MAP2 localises to the dendrites and an initial segment of the axon, promoting Kinesin-3 (KIF1) cargo transport and slowing down Kinesin-1 (KIF5) transport into axons (Gumy et al., 2017).

4.5 INVOLVEMENT IN DISEASE

Mutations identified in Kinesin-3 family members cause hereditary spastic paraplegia (HSP) and related disorders, such as type 2 hereditary sensory and autonomic neuropathy (HSAN2) and progressive encephalopathy with oedema, hypsarrhythmia and optic atrophy (PEHO) syndrome (Gabrych et al., 2019).

HSP is classified into pure (uncomplicated) or complicated forms, based on the presence or absence of additional neurological defects (Harding, 1993, Fink, 2003). It was recently proposed that manifestation of HSP as a pure or complicated form depends on whether the mutation results in gain of function (i.e. motor becomes hyperactive) or loss of function (i.e. motor becomes weak) (Chiba et al., 2019). Since Kinesin-3 motors are implicated in long-distance transport, their disruption would be expected to most affect the longest cells in the body, i.e. sensory and motor neurons in the sciatic nerve. This might explain why the lower limbs are preferentially affected in these patients.

A homozygous mutation in the gene encoding the highly conserved motor domain of KIF1A was identified in a Palestinian family, presenting with early childhood onset resulting in spastic paraplegia (SPG) 30 (Erlich et al., 2011). Additional mutations were shown by exome sequencing along the regulatory and cargo-binding regions in several families to cause HSAN2 (Riviere et al., 2011), SPG 30 (Klebe et al., 2006, Klebe et al., 2012) or PEHO syndrome (Langlois et al., 2016). In cases where the mutation is heterozygous, it manifests as a non-syndromic intellectual disability (mental retardation, autosomal dominant 9: MRD9) with cerebellar atrophy and axonal neuropathy (Hamdan et al., 2011, Ohba et al., 2015, Yoshikawa et al., 2019). Interestingly, mutations T99M and E253K responsible for SPG 30 are also implicated in PEHO syndrome (Samanta and Gokden, 2019, Esmaeeli Nieh et al., 2015, Lee et al., 2015a). *In vitro,* these two mutations resulted in non-motility in gliding assays, failing to localise to peripheral regions of cultured hippocampal neurons and accumulation in proximal axon regions instead (Esmaeeli Nieh et al., 2015, Cheon et al., 2017). Other KIF1A mutations tested *in vitro,* such as V8M, suggest that an over-activation of the motor results in gain-of-function mutations, leading to pure SPG (Chiba et al., 2019).

For KIF1C, mutations identified in two different families resulted in a novel complicated form of SPG 58, which presented with cerebellar ataxia and pyramidal tract dysfunction (Dor et al., 2014, Caballero Oteyza et al., 2014). Additional mutations, causing either KIF1C truncated at the stalk region or nonsense-degradation, have also been identified (Yucel-Yilmaz et al., 2018, Marchionni et al., 2019). A homozygous single nucleotide polymorphism in bovine KIF1C results in loss of protein and causes progressive ataxia in Charolais cattle (Duchesne et al., 2018). Interestingly, this KIF1C mutation is correlated with desirable traits in this cattle breed. A slightly better muscular and skeletal development and higher weight of heterozygote carriers of the KIF1C mutation explains why the incidence of the mutation has been maintained at a high frequency in this breed (Duchesne et al., 2018).

A list of known disease-causing mutations in Kinesin-3s is available in online downloadable material (Table 4.3).

ACKNOWLEDGEMENTS

This work was funded by a Wellcome Trust Investigator Award (200870/Z/16/Z) to A.S.

REFERENCES

Al-Bassam, J., Cui, Y., Klopfenstein, D., Carragher, B. O., Vale, R. D. & Milligan, R. A. 2003. Distinct conformations of the kinesin Unc104 neck regulate a monomer to dimer motor transition. *J Cell Biol*, 163, 743–53.

Atherton, J., Farabella, I., Yu, I. M., Rosenfeld, S. S., Houdusse, A., Topf, M. & Moores, C. A. 2014. Conserved mechanisms of microtubule-stimulated ADP release, ATP binding, and force generation in transport kinesins. *Elife*, 3, e03680.

Bhuwania, R., Castro-Castro, A. & Linder, S. 2014. Microtubule acetylation regulates dynamics of KIF1C-powered vesicles and contact of microtubule plus ends with podosomes. *Eur J Cell Biol*, 93, 424–37.

Blatner, N. R., Wilson, M. I., Lei, C., Hong, W., Murray, D., Williams, R. L. & Cho, W. 2007. The structural basis of novel endosome anchoring activity of KIF16B kinesin. *EMBO J*, 26, 3709–19.

Caballero Oteyza, A., Battaloglu, E., Ocek, L., Lindig, T., Reichbauer, J., Rebelo, A. P., Gonzalez, M. A., Zorlu, Y., Ozes, B., Timmann, D., Bender, B., Woehlke, G., Zuchner, S., Schols, L. & Schule, R. 2014. Motor protein mutations cause a new form of hereditary spastic paraplegia. *Neurology*, 82, 2007–16.

Carleton, M., Mao, M., Biery, M., Warrener, P., Kim, S., Buser, C., Marshall, C. G., Fernandes, C., Annis, J. & Linsley, P. S. 2006. RNA interference-mediated silencing of mitotic kinesin KIF14 disrupts cell cycle progression and induces cytokinesis failure. *Mol Cell Biol*, 26, 3853–63.

Chen, C. W., Peng, Y. F., Yen, Y. C., Bhan, P., Muthaiyan Shanmugam, M., Klopfenstein, D. R. & Wagner, O. I. 2019. Insights on UNC-104-dynein/dynactin interactions and their implications on axonal transport in Caenorhabditis elegans. *J Neurosci Res*, 97, 185–201.

Cheon, C. K., Lim, S. H., Kim, Y. M., Kim, D., Lee, N. Y., Yoon, T. S., Kim, N. S., Kim, E. & Lee, J. R. 2017. Autosomal dominant transmission of complicated hereditary spastic paraplegia due to a dominant negative mutation of KIF1A, SPG30 gene. *Sci Rep*, 7, 12527.

Chiba, K., Min, C., Arai, S., Hashimoto, K., McKenney, R. J. & Niwa, S. 2019. Disease-associated mutations hyperactivate KIF1A motility and anterograde axonal transport of synaptic vesicle precursors. *Proc Natl Acad Sci U S A*, 116(37), 18429–34.

Delevoye, C., Miserey-Lenkei, S., Montagnac, G., Gilles-Marsens, F., Paul-Gilloteaux, P., Giordano, F., Waharte, F., Marks, M. S., Goud, B. & Raposo, G. 2014. Recycling endosome tubule morphogenesis from sorting endosomes requires the kinesin motor KIF13A. *Cell Rep*, 6, 445–54.

Dor, T., Cinnamon, Y., Raymond, L., Shaag, A., Bouslam, N., Bouhouche, A., Gaussen, M., Meyer, V., Durr, A., Brice, A., Benomar, A., Stevanin, G., Schuelke, M. & Edvardson, S. 2014. KIF1C mutations in two families with hereditary spastic paraparesis and cerebellar dysfunction. *J Med Genet*, 51, 137–42.

Dorner, C., Ciossek, T., Muller, S., Moller, P. H., Ullrich, A. & Lammers, R. 1998. Characterization of KIF1C, a new kinesin-like protein involved in vesicle transport from the Golgi apparatus to the endoplasmic reticulum. *J Biol Chem*, 273, 20267–75.

Dorner, C., Ullrich, A., Haring, H. U. & Lammers, R. 1999. The kinesin-like motor protein KIF1C occurs in intact cells as a dimer and associates with proteins of the 14-3-3 family. *J Biol Chem*, 274, 33654–60.

Duchesne, A., Vaiman, A., Frah, M., Floriot, S., Legoueix-Rodriguez, S., Desmazieres, A., Fritz, S., Beauvallet, C., Albaric, O., Venot, E., Bertaud, M., Saintilan, R., Guatteo, R., Esquerre, D., Branchu, J., Fleming, A., Brice, A., Darios, F., Vilotte, J. L., Stevanin, G., Boichard, D. & El Hachimi, K. H. 2018. Progressive ataxia of Charolais cattle highlights a role of KIF1C in sustainable myelination. *PLoS Genet*, 14, e1007550.

Durocher, D. & Jackson, S. P. 2002. The FHA domain. *FEBS Lett*, 513, 58–66.

Efimova, N., Grimaldi, A., Bachmann, A., Frye, K., Zhu, X., Feoktistov, A., Straube, A. & KAVERINA, I. 2014. Podosome-regulating kinesin KIF1C translocates to the cell periphery in a CLASP-dependent manner. *J Cell Sci*, 127, 5179–88.

Erlich, Y., Edvardson, S., Hodges, E., Zenvirt, S., Thekkat, P., Shaag, A., Dor, T., Hannon, G. J. & Elpeleg, O. 2011. Exome sequencing and disease-network analysis of a single family implicate a mutation in KIF1A in hereditary spastic paraparesis. *Genome Res*, 21, 658–64.

Esmaeeli Nieh, S., Madou, M. R., Sirajuddin, M., Fregeau, B., McKnight, D., Lexa, K., Strober, J., Spaeth, C., Hallinan, B. E., Smaoui, N., Pappas, J. G., Burrow, T. A., McDonald, M. T., Latibashvili, M., Leshinsky-Silver, E., Lev, D., Blumkin, L., Vale, R. D., Barkovich, A. J. & Sherr, E. H. 2015. De novo mutations in KIF1A cause progressive encephalopathy and brain atrophy. *Ann Clin Transl Neurol*, 2, 623–35.

Etoh, K. & Fukuda, M. 2019. Rab10 regulates tubular endosome formation through KIF13A and KIF13B motors. *J Cell Sci*, 132(5), pii: jcs226977.

Farkhondeh, A., Niwa, S., Takei, Y. & Hirokawa, N. 2015. Characterizing KIF16B in neurons reveals a novel intramolecular "stalk inhibition" mechanism that regulates its capacity to potentiate the selective somatodendritic localization of early endosomes. *J Neurosci*, 35, 5067–86.

Fink, J. K. 2003. Advances in the hereditary spastic paraplegias. *Exp Neurol*, 184 Suppl 1, S106–10.

Gabrych, D. R., Lau, V. Z., Niwa, S. & Silverman, M. A. 2019. Going too far is the same as falling short(dagger): Kinesin-3 family members in hereditary spastic paraplegia. *Front Cell Neurosci*, 13, 419.

Goodsell, D. S., Autin, L. & Olson, A. J. 2019. Illustrate: Software for biomolecular Illustration. *Structure*, 27, 1716–20 e1.

Guardia, C. M., Farias, G. G., Jia, R., Pu, J. & Bonifacino, J. S. 2016. BORC functions upstream of kinesins 1 and 3 to coordinate regional movement of lysosomes along different microtubule tracks. *Cell Rep*, 17, 1950–61.

Gumy, L. F., Katrukha, E. A., Grigoriev, I., Jaarsma, D., Kapitein, L. C., Akhmanova, A. & Hoogenraad, C. C. 2017. MAP2 defines a pre-axonal filtering zone to regulate KIF1- versus KIF5-dependent cargo transport in sensory neurons. *Neuron*, 94, 347–62 e7.

Hamdan, F. F., Gauthier, J., Araki, Y., Lin, D. T., Yoshizawa, Y., Higashi, K., Park, A. R., Spiegelman, D., Dobrzeniecka, S., Piton, A., Tomitori, H., Daoud, H., Massicotte, C., Henrion, E., Diallo, O., Group, S. D., Shekarabi, M., Marineau, C., Shevell, M., Maranda, B., Mitchell, G., Nadeau, A., D'Anjou, G., Vanasse, M., Srour, M., Lafreniere, R. G., Drapeau, P., Lacaille, J. C., Kim, E., Lee, J. R., Igarashi, K., Huganir, R. L.,

Rouleau, G. A. & Michaud, J. L. 2011. Excess of de novo deleterious mutations in genes associated with glutamatergic systems in nonsyndromic intellectual disability. *Am J Hum Genet*, 88, 306–16.

Hammond, J. W., Cai, D., Blasius, T. L., Li, Z., Jiang, Y., Jih, G. T., Meyhofer, E. & Verhey, K. J. 2009. Mammalian Kinesin-3 motors are dimeric in vivo and move by processive motility upon release of autoinhibition. *PLoS Biol*, 7, e72.

Harding, A. E. 1993. Hereditary spastic paraplegias. *Semin Neurol*, 13, 333–6.

Hoepfner, S., Severin, F., Cabezas, A., Habermann, B., Runge, A., Gillooly, D., Stenmark, H. & Zerial, M. 2005. Modulation of receptor recycling and degradation by the endosomal kinesin KIF16B. *Cell*, 121, 437–50.

Huckaba, T. M., Gennerich, A., Wilhelm, J. E., Chishti, A. H. & Vale, R. D. 2011. Kinesin-73 is a processive motor that localizes to Rab5-containing organelles. *J Biol Chem*, 286, 7457–67.

Huo, L., Yue, Y., Ren, J., Yu, J., Liu, J., Yu, Y., Ye, F., Xu, T., Zhang, M. & Feng, W. 2012. The CC1-FHA tandem as a central hub for controlling the dimerization and activation of kinesin-3 KIF1A. *Structure*, 20, 1550–61.

Ikegami, K., Heier, R. L., Taruishi, M., Takagi, H., Mukai, M., Shimma, S., Taira, S., Hatanaka, K., Morone, N., Yao, I., Campbell, P. K., Yuasa, S., Janke, C., Macgregor, G. R. & Setou, M. 2007. Loss of alpha-tubulin polyglutamylation in ROSA22 mice is associated with abnormal targeting of KIF1A and modulated synaptic function. *Proc Natl Acad Sci U S A*, 104, 3213–18.

Janke, C. 2014. The tubulin code: Molecular components, readout mechanisms, and functions. *J Cell Biol*, 206, 461–72.

Kendrick, A. A., Dickey, A. M., Redwine, W. B., Tran, P. T., Vaites, L. P., Dzieciatkowska, M., Harper, J. W. & Reck-Peterson, S. L. 2019. Hook3 is a scaffold for the opposite-polarity microtubule-based motors cytoplasmic dynein-1 and KIF1C. *J Cell Biol*, 218(9), 2982–3001.

Klebe, S., Azzedine, H., Durr, A., Bastien, P., Bouslam, N., Elleuch, N., Forlani, S., Charon, C., Koenig, M., Melki, J., Brice, A. & Stevanin, G. 2006. Autosomal recessive spastic paraplegia (SPG30) with mild ataxia and sensory neuropathy maps to chromosome 2q37.3. *Brain*, 129, 1456–62.

Klebe, S., Lossos, A., Azzedine, H., Mundwiller, E., Sheffer, R., Gaussen, M., Marelli, C., Nawara, M., Carpentier, W., Meyer, V., Rastetter, A., Martin, E., Bouteiller, D., Orlando, L., Gyapay, G., El-Hachimi, K. H., Zimmerman, B., Gamliel, M., Misk, A., Lerer, I., Brice, A., Durr, A. & Stevanin, G. 2012. KIF1A missense mutations in SPG30, an autosomal recessive spastic paraplegia: Distinct phenotypes according to the nature of the mutations. *Eur J Hum Genet*, 20, 645–9.

Kopp, P., Lammers, R., Aepfelbacher, M., Woehlke, G., Rudel, T., Machuy, N., Steffen, W. & Linder, S. 2006. The kinesin KIF1C and microtubule plus ends regulate podosome dynamics in macrophages. *Mol Biol Cell*, 17, 2811–23.

Langlois, S., Tarailo-Graovac, M., Sayson, B., Drogemoller, B., Swenerton, A., Ross, C. J., Wasserman, W. W. & Van Karnebeek, C. D. 2016. De novo dominant variants affecting the motor domain of KIF1A are a cause of PEHO syndrome. *Eur J Hum Genet*, 24, 949–53.

Lee, J. R., Srour, M., Kim, D., Hamdan, F. F., Lim, S. H., Brunel-Guitton, C., Decarie, J. C., Rossignol, E., Mitchell, G. A., Schreiber, A., Moran, R., Van Haren, K., Richardson, R., Nicolai, J., Oberndorff, K. M., Wagner, J. D., Boycott, K. M., Rahikkala, E., Junna, N., Tyynismaa, H., Cuppen, I., Verbeek, N. E., Stumpel, C. T., Willemsen, M. A., De Munnik, S. A., Rouleau, G. A., Kim, E., Kamsteeg, E. J., Kleefstra, T. & Michaud, J. L. 2015a. De novo mutations in the motor domain of KIF1A cause cognitive impairment, spastic paraparesis, axonal neuropathy, and cerebellar atrophy. *Hum Mutat*, 36, 69–78.

Lee, P. L., Ohlson, M. B. & Pfeffer, S. R. 2015b. Rab6 regulation of the kinesin family KIF1C motor domain contributes to Golgi tethering. *Elife*, 4, e06029.

Lessard, D. V., Zinder, O. J., Hotta, T., Verhey, K. J., Ohi, R. & Berger, C. L. 2019. Polyglutamylation of tubulin's C-terminal tail controls pausing and motility of kinesin-3 family member KIF1A. *J Biol Chem*, 294(16), 6353–63.

Lipka, J., Kapitein, L. C., Jaworski, J. & Hoogenraad, C. C. 2016. Microtubule-binding protein doublecortin-like kinase 1 (DCLK1) guides kinesin-3-mediated cargo transport to dendrites. *EMBO J*, 35, 302–18.

Magiera, M. M. & Janke, C. 2014. Post-translational modifications of tubulin. *Curr Biol*, 24, R351–4.

Marchionni, E., Meneret, A., Keren, B., Melki, J., Denier, C., Durr, A., Apartis, E., Boespflug-Tanguy, O. & Mochel, F. 2019. KIF1C Variants are associated with hypomyelination, ataxia, tremor, and dystonia in fraternal twins. *Tremor Other Hyperkinet Mov (N Y)*, 9.

Matsushita, M., Yamamoto, R., Mitsui, K. & Kanazawa, H. 2009. Altered motor activity of alternative splice variants of the mammalian kinesin-3 protein KIF1B. *Traffic*, 10, 1647–54.

Miki, H., Okada, Y. & Hirokawa, N. 2005. Analysis of the kinesin superfamily: Insights into structure and function. *Trends Cell Biol*, 15, 467–76.

Mills, J., Hanada, T., Hase, Y., Liscum, L. & Chishti, A. H. 2019. LDL receptor related protein 1 requires the I3 domain of discs-large homolog 1/DLG1 for interaction with the kinesin motor protein KIF13B. *Biochim Biophys Acta Mol Cell Res*, 1866, 118552.

Monroy, B. Y., Sawyer, D. L., Ackermann, B. E., Borden, M. M., Tan, T. C. & Ori-McKenney, K. M. 2018. Competition between microtubule-associated proteins directs motor transport. *Nat Commun*, 9, 1487.

Monroy, B. Y., Tan, T. C., Oclaman, J. M., Han, J. S., Simo, S., Nowakowski, D. W., McKenney, R. J. & Ori-McKenney, K. M. 2020. A combinatorial MAP code dictates polarized microtubule transport. *Dev Cell*, doi: 10.1016/j.devcel.2020.01.029. S1534-5807(20)30061-7.

Nakagawa, T., Setou, M., Seog, D., Ogasawara, K., Dohmae, N., Takio, K. & Hirokawa, N. 2000. A novel motor, KIF13A, transports mannose-6-phosphate receptor to plasma membrane through direct interaction with AP-1 complex. *Cell*, 103, 569–81.

Nonet, M. L. 1999. Visualization of synaptic specializations in live C. elegans with synaptic vesicle protein-GFP fusions. *J Neurosci Methods*, 89, 33–40.

O'Hagan, R., Piasecki, B. P., Silva, M., Phirke, P., Nguyen, K. C., Hall, D. H., Swoboda, P. & Barr, M. M. 2011. The tubulin deglutamylase CCPP-1 regulates the function and stability of sensory cilia in C. elegans. *Curr Biol*, 21, 1685–94.

Ohba, C., Haginoya, K., Osaka, H., Kubota, K., Ishiyama, A., Hiraide, T., Komaki, H., Sasaki, M., Miyatake, S., Nakashima, M., Tsurusaki, Y., Miyake, N., Tanaka, F., Saitsu, H. & Matsumoto, N. 2015. De novo KIF1A mutations cause intellectual deficit, cerebellar atrophy, lower limb spasticity and visual disturbance. *J Hum Genet*, 60, 739–42.

Okada, Y., Higuchi, H. & Hirokawa, N. 2003. Processivity of the single-headed kinesin KIF1A through biased binding to tubulin. *Nature*, 424, 574–7.

Okada, Y. & Hirokawa, N. 1999. A processive single-headed motor: Kinesin superfamily protein KIF1A. *Science*, 283, 1152–7.

Okada, Y. & Hirokawa, N. 2000. Mechanism of the single-headed processivity: Diffusional anchoring between the K-loop of kinesin and the C terminus of tubulin. *Proc Natl Acad Sci U S A*, 97, 640–5.

Okada, Y., Yamazaki, H., Sekine-Aizawa, Y. & Hirokawa, N. 1995. The neuron-specific kinesin superfamily protein KIF1A is a unique monomeric motor for anterograde axonal transport of synaptic vesicle precursors. *Cell*, 81, 769–80.

Oriola, D. & Casademunt, J. 2013. Cooperative force generation of KIF1A Brownian motors. *Phys Rev Lett*, 111, 048103.

Peckham, M. 2011. Coiled coils and SAH domains in cytoskeletal molecular motors. *Biochem Soc Trans*, 39, 1142–8.

Redwine, W. B., Desantis, M. E., Hollyer, I., Htet, Z. M., Tran, P. T., Swanson, S. K., Florens, L., Washburn, M. P. & Reck-Peterson, S. L. 2017. The human cytoplasmic dynein inter-actome reveals novel activators of motility. *Elife*, 6, e28257.

Ren, J., Huo, L., Wang, W., Zhang, Y., Li, W., Lou, J., Xu, T. & Feng, W. 2016. Structural cor-relation of the neck coil with the coiled-coil (CC1)-forkhead-associated (FHA) tandem for active kinesin-3 KIF13A. *J Biol Chem*, 291, 3581–94.

Ren, J., Wang, S., Chen, H., Wang, W., Huo, L. & Feng, W. 2018. Coiled-coil 1-mediated fastening of the neck and motor domains for kinesin-3 autoinhibition. *Proc Natl Acad Sci U S A*, 115, E11933–42.

Riviere, J. B., Ramalingam, S., Lavastre, V., Shekarabi, M., Holbert, S., Lafontaine, J., Srour, M., Merner, N., Rochefort, D., Hince, P., Gaudet, R., Mes-Masson, A. M., Baets, J., Houlden, H., Brais, B., Nicholson, G. A., Van Esch, H., Nafissi, S., De Jonghe, P., Reilly, M. M., Timmerman, V., Dion, P. A. & Rouleau, G. A. 2011. KIF1A, an axonal transporter of synaptic vesicles, is mutated in hereditary sensory and autonomic neu-ropathy type 2. *Am J Hum Genet*, 89, 219–30.

Rogers, K. R., Weiss, S., Crevel, I., Brophy, P. J., Geeves, M. & Cross, R. 2001. KIF1D is a fast non-processive kinesin that demonstrates novel K-loop-dependent mechanochemistry. *EMBO J*, 20, 5101–13.

Sagona, A. P., Nezis, I. P., Pedersen, N. M., Liestol, K., Poulton, J., Rusten, T. E., Skotheim, R. I., Raiborg, C. & Stenmark, H. 2010. PtdIns(3)P controls cytokinesis through KIF13A-mediated recruitment of FYVE-CENT to the midbody. *Nat Cell Biol*, 12, 362–71.

Samanta, D. & Gokden, M. 2019. PEHO syndrome: KIF1A mutation and decreased activity of mitochondrial respiratory chain complex. *J Clin Neurosci*, 61, 298–301.

Scarabelli, G., Soppina, V., Yao, X. Q., Atherton, J., Moores, C. A., Verhey, K. J. & Grant, B. J. 2015. Mapping the processivity determinants of the Kinesin-3 motor domain. *Biophys J*, 109, 1537–40.

Schlager, M. A., Kapitein, L. C., Grigoriev, I., Burzynski, G. M., Wulf, P. S., Keijzer, N., De Graaff, E., Fukuda, M., Shepherd, I. T., Akhmanova, A. & Hoogenraad, C. C. 2010. Pericentrosomal targeting of Rab6 secretory vesicles by Bicaudal-D-related protein 1 (BICDR-1) regulates neuritogenesis. *EMBO J*, 29, 1637–51.

Seidel, C., Zekert, N. & Fischer, R. 2012. The *Aspergillus nidulans* kinesin-3 tail is necessary and sufficient to recognize modified microtubules. *PLoS One*, 7, e30976.

Siddiqui, N. & Straube, A. 2017. Intracellular cargo transport by kinesin-3 motors. *Biochemistry (Mosc)*, 82, 803–15.

Siddiqui, N., Zwetsloot, A. J., Bachmann, A., Roth, D., Hussain, H., Brandt, J., Kaverina, I. & Straube, A. 2019. PTPN21 and Hook3 relieve KIF1C autoinhibition and activate intracellular transport. *Nat Commun*, 10, 2693.

Soppina, V., Norris, S. R., Dizaji, A. S., Kortus, M., Veatch, S., Peckham, M. & Verhey, K. J. 2014. Dimerization of mammalian kinesin-3 motors results in superprocessive motion. *Proc Natl Acad Sci U S A*, 111, 5562–7.

Soppina, V. & Verhey, K. J. 2014. The family-specific K-loop influences the microtubule on-rate but not the superprocessivity of kinesin-3 motors. *Mol Biol Cell*, 25, 2161–70.

Theisen, U., Straube, E. & Straube, A. 2012. Directional persistence of migrating cells requires Kif1C-mediated stabilization of trailing adhesions. *Dev Cell*, 23, 1153–66.

Tomishige, M., Klopfenstein, D. R. & Vale, R. D. 2002. Conversion of Unc104/KIF1A kine-sin into a processive motor after dimerization. *Science*, 297, 2263–7.

Torres, J. Z., Summers, M. K., Peterson, D., Brauer, M. J., Lee, J., Senese, S., Gholkar, A. A., Lo, Y. C., Lei, X., Jung, K., Anderson, D. C., Davis, D. P., Belmont, L. & Jackson, P. K. 2011. The STARD9/Kif16a kinesin associates with mitotic microtubules and regulates spindle pole assembly. *Cell*, 147, 1309–23.

Ueno, H., Huang, X., Tanaka, Y. & Hirokawa, N. 2011. KIF16B/Rab14 molecular motor complex is critical for early embryonic development by transporting FGF receptor. *Dev Cell*, 20, 60–71.

Urnavicius, L., Lau, C. K., Elshenawy, M. M., Morales-rios, E., Motz, C., Yildiz, A. & Carter, A. P. 2018. Cryo-EM shows how dynactin recruits two dyneins for faster movement. *Nature*, 554, 202–6.

Watters, J. W., Dewar, K., Lehoczky, J., Boyartchuk, V. & Dietrich, W. F. 2001. Kif1C, a kinesin-like motor protein, mediates mouse macrophage resistance to anthrax lethal factor. *Curr Biol*, 11, 1503–11.

Wedlich-Soldner, R. 2002. A balance of KIF1A-like kinesin and dynein organizes early endosomes in the fungus Ustilago maydis. *EMBO J*, 21, 2946–57.

Westerholm-Parvinen, A., Vernos, I. & Serrano, L. 2000. Kinesin subfamily UNC104 contains a FHA domain: Boundaries and physicochemical characterization. *FEBS Lett*, 486, 285–90.

Wickstead, B. & K. Gull. 2006. A "holistic" kinesin phylogeny reveals new kinesin families and predicts protein functions. *Mol Biol Cell*, 17(4), 1734–43. doi:10.1091/mbc. e05-11-1090.

Xing, B. M., Yang, Y. R., Du, J. X., Chen, H. J., Qi, C., Huang, Z. H., Zhang, Y. & Wang, Y. 2012. Cyclin-dependent kinase 5 controls TRPV1 membrane trafficking and the heat sensitivity of nociceptors through KIF13B. *J Neurosci*, 32, 14709–21.

Xue, X., Jaulin, F., Espenel, C. & Kreitzer, G. 2010. PH-domain-dependent selective transport of p75 by kinesin-3 family motors in non-polarized MDCK cells. *J Cell Sci*, 123, 1732–41.

Yamada, K. H., Hanada, T. & Chishti, A. H. 2007. The effector domain of human Dlg tumor suppressor acts as a switch that relieves autoinhibition of kinesin-3 motor GAKIN/KIF13B. *Biochemistry*, 46, 10039–45.

Yoshikawa, K., Kuwahara, M., Saigoh, K., Ishiura, H., Yamagishi, Y., Hamano, Y., Samukawa, M., Suzuki, H., Hirano, M., Mitsui, Y., Tsuji, S. & Kusunoki, S. 2019. The novel de novo mutation of KIF1A gene as the cause for Spastic paraplegia 30 in a Japanese case. *eNeurologicalSci*, 14, 34–7.

Yoshimura, Y., Terabayashi, T. & Miki, H. 2010. Par1b/MARK2 phosphorylates kinesin-like motor protein GAKIN/KIF13B to regulate axon formation. *Mol Cell Biol*, 30, 2206–19.

Yucel-Yilmaz, D., Yucesan, E., Yalnizoglu, D., Oguz, K. K., Sagiroglu, M. S., Ozbek, U., Serdaroglu, E., Bilgic, B., Erdem, S., Iseri, S. A. U., Hanagasi, H., Gurvit, H., Ozgul, R. K. & Dursun, A. 2018. Clinical phenotype of hereditary spastic paraplegia due to KIF1C gene mutations across life span. *Brain Dev*, 40, 458–64.

Yue, Y., Sheng, Y., Zhang, H. N., Yu, Y., Huo, L., Feng, W. & Xu, T. 2013. The CC1-FHA dimer is essential for KIF1A-mediated axonal transport of synaptic vesicles in C. elegans. *Biochem Biophys Res Commun*, 435, 441–6.

Zekert, N. & Fischer, R. 2009. The Aspergillus nidulans kinesin-3 UncA motor moves vesicles along a subpopulation of microtubules. *Mol Biol Cell*, 20, 673–84.

Zerial, M. & McBride, H. 2001. Rab proteins as membrane organizers. *Nat Rev Mol Cell Biol*, 2, 107–17.

5 The Kinesin-4 Family

Claire T. Friel

CONTENTS

Members of the Kinesin-4 family, which incorporates the previously identified Kinesin-10 family, have the ability to alter microtubule dynamics. They display a range of modes of movement upon microtubules, from diffusive motion with no consistent direction, to plus-end-directed movement with varying degrees of processivity.

5.1 EXAMPLE FAMILY MEMBERS

Mammalian: KIF4A, KIF4B, KIF7, KIF21A, KIF21B, KIF27
Drosophila melanogaster: KLP3A, KLP31E
Caenorhabditis elegans: KLP-12, KLP-19
Xenopus laevis: XKLP1
Arabidopsis thaliana: FRA1

5.2 STRUCTURAL INFORMATION

The motor domain of members of the Kinesin-4 family is located N-terminally in the primary sequence and is immediately followed by a short neck-linker region (Figure 5.1). This is followed by a discontinuous α-helical region, predicted to form a coiled-coil facilitating oligomerisation, and a C-terminal globular tail domain. This domain architecture is typical across the Kinesin-4 family. A crystal structure exists

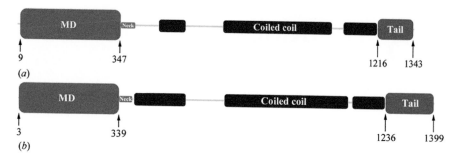

FIGURE 5.1 Typical domain layout for the Kinesin-4 family members (A) KIF7 and (B) KIF27. The globular motor domain (MD) and short neck linker (Neck) are followed by a region predicted to form a discontinuous coiled-coil domain (Coiled coil) and a C-terminal globular tail domain (Tail). Taken from Klejnot and Kozielski (2012).

for the Kinesin-4, KIF7, motor domain which contains the eight-stranded β-sheet core with three major α-helices on either side characteristic of the kinesin superfamily (Klejnot and Kozielski 2012).

Many members of the Kinesin-4 family interact with chromosomes, and the family at one time was considered to be a family of chromokinesins (Mazumdar and Misteli 2005). The interaction with DNA is mediated via the C-terminal part of the sequence. Human KIF4, which is highly conserved across different species (Vernos et al. 1995, Williams et al. 1995, Oh et al. 2000, Powers et al. 2004), contains two conserved motifs critical for binding to chromatin (Wu and Chen 2008). A ZIP1/ZBZ leucine zipper motif is found within the coiled-coil region and a cysteine-rich (CR) motif is located in the globular tail domain. The interaction with DNA and DNA-related proteins, such as condensin (Takahashi, Wakai, and Hirota 2016) and PRC1 (Kurasawa et al. 2004), is mediated via these motifs, which are essential for chromosome condensation, DNA repair and DNA replication (Sekine et al. 1994, Wu et al. 2008).

5.3 FUNCTIONAL PROPERTIES

5.3.1 MICROTUBULE MOTILITY

Members of the Kinesin-4 family display a range of modes of interaction with microtubules, although all members studied to date share the ability to alter microtubule dynamics. Many members of the family display motility directed towards the microtubule plus-end, with varying degrees of processivity. The processivity of the motor domain is modulated by additional microtubule-binding sites within the kinesin or by interaction with an external partner with microtubule-binding capabilities.

The mammalian Kinesin-4, KIF4, shows weakly processive, plus-end-directed motility and microtubule-activated ATP turnover (Sekine et al. 1994, Subramanian et al. 2013). The processivity of motility is increased by interaction with the mitotic spindle-associated microtubule-bundling protein, PRC1 (Subramanian et al. 2013). The Kinesin-4, KIF21B, displays plus-end-directed motility with greater processivity than KIF4 (van Riel et al. 2017). This increased processivity results from an additional microtubule-binding region located in the tail, which has a similar effect as the binding of PRC1 to KIF4. Truncated KIF21B, lacking the tail, displays short plus-end-directed runs, like KIF4 in the absence of PRC1. A truncated form of the cilium-associated Kinesin-4, KIF7, lacking the C-terminal region, displays a nucleotide-dependent diffusive interaction with microtubules lacking directed motility (He et al. 2014).

A truncation of the *Xenopus* Kinesin-4, XKLP1, lacking the C-terminal end, also shows weakly processive plus-end-directed movement (Bringmann et al. 2004). However, full-length XKLP1 in the presence of PRC1 is found to accumulate at the microtubule plus-ends, due to long-distance processive motility (Bieling, Telley, and Surrey 2010).

Measurements of motility of C-terminally truncated dimeric versions of Kinesin-4 from *D. melanogaster*, KLP31E, and *C. elegans*, KLP12, on microtubules shows that all of these are plus-end-directed, moderate-velocity kinesins. The same study showed that human KIF27 is a slow plus-end-directed motor and supports the finding that human KIF7 shows no directed motility (Yue et al. 2018).

5.3.2 REGULATION OF MICROTUBULE DYNAMICS

Whether or not they possess directed motility, all members of the Kinesin-4 family studied to date have the ability to alter microtubule dynamics. A study of the impact of an N-terminal fragment of XKLP1 on the dynamics of microtubule asters showed that growth velocity was slowed and eventually completely inhibited as the concentration of XKLP1 increased (Bringmann et al. 2004). The presence of XKLP1 also slowed and eventually completely inhibited microtubule depolymerisation. This effect on microtubule dynamics could be reversed by washing out XKLP1 and did not require ATP turnover. The same effect is seen for full-length XKLP1, which inhibits turnover of tubulin at growing microtubule ends (Bieling, Telley, and Surrey 2010).

KIF7, a Kinesin-4 which localises to the microtubule plus-end at the tip of the cilium, has been shown to reduce the rate of microtubule growth and increase the frequency of catastrophe (He et al. 2014). In other studies, both KIF7 and KIF27 have been shown to inhibit microtubule growth (Yue et al. 2018). The mechanism by which KIF7 recognises the microtubule end is via preferential interaction with GTP-tubulin, which is found at the growing microtubule tip, relative to the GDP-tubulin of the microtubule lattice (Jiang et al. 2019).

The effect of KIF4A on microtubule dynamics is regulated by Aurora B kinase. Phosphorylation of KIF4A by Aurora B increases its microtubule-stimulated rate of ATP turnover and promotes interaction with PRC1 (Nunes Bastos et al. 2013). In the presence of phosphorylated KIF4A, microtubules grow more slowly and show long pauses in growth compared to phospho-null mutants.

5.4 PHYSIOLOGICAL ROLE

The mammalian Kinesin-4, KIF4, was originally discovered in the murine central nervous system and is strongly expressed in juvenile brain tissue (Sekine et al. 1994, Aizawa et al. 1992). Although some studies do not differentiate between KIF4 subtypes, referring only to KIF4, in humans KIF4A and KIF4B are shown to be two closely related but distinct kinesins (Oh et al. 2000, Ha et al. 2000).

5.4.1 THROUGHOUT THE CELL CYCLE

KIF4 interacts with chromatin (Mazumdar, Sundareshan, and Misteli 2004, Samejima et al. 2012), and is involved in setting the length of the mitotic spindle (Hu et al. 2011) and in regulating cytokinesis (Lee and Kim 2004).

KIF4 interacts with DNA and DNA-associated proteins to maintain chromatin and chromosome structure throughout the cell cycle (Mazumdar, Sung, and Misteli 2011, Mitchison et al. 2013, Camlin, McLaughlin, and Holt 2017). During interphase, the majority of KIF4 is located in the nucleus, where it functions as a structural component of chromatin (Mazumdar, Sundareshan, and Misteli 2004, Mazumdar, Sung, and Misteli 2011). After disruption of the nuclear membrane, KIF4A interacts with condensin I and localises along the long axis of chromosomes (Mazumdar, Sundareshan, and Misteli 2004, Samejima et al. 2012, Takahashi, Wakai, and Hirota 2016). RNA interference-mediated knockdown of KIF4A or 4B results in an abnormally elongated mitotic spindle and multinucleated cells, as cells fail to complete cytokinesis (Zhu et al. 2005, Wandke et al. 2012).

Early in mitosis, a proportion of KIF4 moves to the plus-ends of non-kinetochore microtubules to regulate microtubule growth, and the absence of KIF4 results in abnormal elongation of the midzone and unfocused overlap regions (Hu et al. 2011). The *Drosophila* Kinesin-4, KLP3A, also associates with chromosomes and the mitotic spindle to organise bundles of interpolar microtubules (Kwon et al. 2004). *Xenopus* Kinesin-4, XKLP1, is recruited to antiparallel microtubules by the microtubule-bundling protein PRC1, where it selectively inhibits the growth of overlapping microtubules (Bieling, Telley, and Surrey 2010). This is likely also the mechanism by which KIF4 regulates spindle midzone length. Human KIF4 binds to PRC1; in KIF4-deficient cells, PRC1 fails to concentrate at the midzone, and both midzone formation and cytokinesis are inhibited (Zhu et al. 2005, Kurasawa et al. 2004). During cytokinesis, KIF4-mediated regulation of midzone length is required to focus the cleavage furrow (Hu et al. 2011).

KIF4 appears to play similar roles in meiosis, localising to chromosomes throughout metaphase and transitioning to the midzone during anaphase (Heath and Wignall 2019). In mouse oocytes, KIF4 knockdown results in defective midzone formation and elongated spindles (Heath and Wignall 2019, Camlin, McLaughlin, and Holt 2017). KLP-19 in *C. elegans* oocytes associates with chromatids during meiosis and plays a role in chromosome congression (Wignall and Villeneuve 2009).

5.4.2 DNA DAMAGE RESPONSE

KIF4A has been shown to bind to BRCA2 via its C-terminal cargo-binding domain (Wu et al. 2008). BRCA2 plays an essential role in the regulation of Rad51-driven DNA recombinase activity. When DNA damage is caused by laser micro-irradiation, KIF4A is rapidly recruited to sites of damage. When expression of KIF4A is knocked down, the formation of Rad51 DNA damage foci is impaired and homologous recombination is significantly reduced. Cells in which KIF4A is depleted become highly sensitive to ionising radiation (Wu et al. 2008).

This role for KIF4A in DNA repair underlies its effect as a modulator of sensitivity to the anticancer therapy cisplatin, which acts, among other mechanisms, by inducing DNA double-strand breaks. Cisplatin treatment is shown to stimulate expression of KIF4A in non-small-cell lung cancer cell lines, with depletion of KIF4A enhancing sensitivity to the cytotoxic effect of cisplatin via inhibition of the formation of BRCA2/Rad51 DNA repair foci (Wan et al. 2019).

5.4.3 NERVE CELL DEVELOPMENT

The mammalian Kinesin-4, KIF21B, plays a role in the regulation of both transport and microtubule dynamics in dendrites (Ghiretti et al. 2016). These functions are independent of one another and neuronal activity is found to enhance the transport function of KIF21B at the expense of its microtubule-remodelling function. The closely related KIF21A accumulates in growth cones of axons and is recruited to the cortex by interaction with the protein KANK1 (van der Vaart et al. 2013). KIF21A suppresses microtubule dynamics and is involved in organising microtubule arrays at the cell edge.

KIF4A is expressed in juvenile neurons and is involved in activity-dependent neuronal survival, in which unnecessary neurons are eliminated during brain development. Activity-dependent prevention of apoptosis of juvenile nerve cells is mediated by suppression of the activity of a nuclear enzyme, PARP-1, via interaction with the C-terminal domain of KIF4 (Midorikawa, Takei, and Hirokawa 2006). KIF4 has also been shown to be involved in trafficking of ribosomal components in axons and of cell adhesion molecules implicated in axon elongation (Bisbal et al. 2009, Heintz et al. 2014, Peretti et al. 2000).

5.4.4 CILIA

The human Kinesin-4, KIF7, localises to the tips of primary cilia and plays a critical role in correct formation of cilium structure. Primary cilia, formed by cultured fibroblasts with a mutated form of KIF7, are longer and less stable than those formed by wild-type cells (He et al. 2014). KIF7 does not perform a transport function in cilia but rather accumulates at the plus-ends of axonemal microtubules and regulates cilium length via its impact on microtubule dynamics. The Kinesin-4, KIF27, is also associated with cilia, and evidence suggests a role in the formation and function of motile cilia (Vogel et al. 2012, Wilson et al. 2009).

KIF7 plays a key role in the Hedgehog signalling pathway, with its function suggested to be control of cilium tip architecture to create a single ciliary tip compartment from which the activity of other Hedgehog pathway components can be correctly regulated (Cheung et al. 2009, Endoh-Yamagami et al. 2009).

5.5 INVOLVEMENT IN DISEASE

KIF4 is abnormally expressed in a variety of cancers and plays crucial roles in the progression of cancers, including promotion of drug resistance or inhibition of apoptosis (Sheng et al. 2018). The mechanism of action of KIF4 in influencing cancer progression is often unclear and the impact of KIF4 overexpression varies between cancer types. For example, overexpression of KIF4 inhibits proliferation of gastric cancer cell lines (Gao et al. 2011) but enhances both proliferation and invasiveness of liver cancer cell lines (Hou et al. 2017). Two different mechanisms have been described for the impact of KIF4 expression on drug resistance in cancers. Overexpression of KIF4A results in resistance to doxorubin in breast cancer cell lines via suppression of apoptosis by inhibiting the activity of PARP-1 (Wang et al. 2014), whilst resistance of lung cancer cell lines to cisplatin is mediated via upregulation of DNA repair mechanisms by KIF4A (Wan et al. 2019). Bioinformatic analysis of available data on the expression of KIF4A in breast cancer suggests that KIF4A may be a strong prognostic biomarker for breast cancer and a promising therapeutic target (Xue et al. 2018).

Due to its role in neural cell development, certain mutations of KIF4A have pathogenic consequences, such as intellectual disability caused by an imbalance in excitatory and inhibitory synaptic activity (Willemsen et al. 2014). Also associated with a role in neural development, mutations in KIF21A are linked to both familial and sporadic classes of congenital fibrosis of the extraocular muscles type 1 (CFEOM1), a disorder associated with defects of the oculomotor nerve (Yamada et al. 2003). CFEOM1-associated mutations, found in a primary mutational hotspot within the stalk domain, have been shown to relieve autoinhibition of the KIF21A motor domain and result in aberrant axon morphology and accumulation of KIF21A at axonal growth cones (van der Vaart et al. 2013).

KIF4 is also shown to be involved in viral replication via the transport of group-specific antigen (Gag) polyproteins to the plasma membrane (Kim et al. 1998, Martinez et al. 2008). Microarray data suggests a role for KIF4 in the growth and survival of macrophages, possibly via involvement in lipid metabolism (Luan et al. 2019, Xu et al. 2017).

Due to their involvement in cilium development and function, mutations to the Kinesin-4 proteins KIF7 and KIF27 result in a number of ciliopathies, such as Joubert syndrome (Dafinger et al. 2011). Mutations to KIF7 result in developmental disorders consistent with its role in Hedgehog signalling and are found in individuals with hydrolethalus and acrocallosal syndromes (Putoux et al. 2012, Putoux et al. 2011). Mice with KIF27 knocked out do not survive beyond eight weeks after birth and exhibit hydrocephalus (Vogel et al. 2012).

REFERENCES

Aizawa, H., Y. Sekine, R. Takemura, Z. Zhang, M. Nangaku, and N. Hirokawa. 1992. "Kinesin family in murine central nervous system." *J Cell Biol* 119 (5):1287–96. doi:10.1083/jcb.119.5.1287.

Bieling, P., I. A. Telley, and T. Surrey. 2010. "A minimal midzone protein module controls formation and length of antiparallel microtubule overlaps." *Cell* 142 (3):420–32. doi:10.1016/j.cell.2010.06.033.

Bisbal, M., J. Wojnacki, D. Peretti, A. Ropolo, J. Sesma, I. Jausoro, and A. Caceres. 2009. "KIF4 mediates anterograde translocation and positioning of ribosomal constituents to axons." *J Biol Chem* 284 (14):9489–97. doi:10.1074/jbc.M808586200.

Bringmann, H., G. Skiniotis, A. Spilker, S. Kandels-Lewis, I. Vernos, and T. Surrey. 2004. "A kinesin-like motor inhibits microtubule dynamic instability." *Science* 303 (5663):1519–22. doi:10.1126/science.1094838.

Camlin, N. J., E. A. McLaughlin, and J. E. Holt. 2017. "Kif4 is essential for mouse oocyte meiosis." *PLoS One* 12 (1):e0170650. doi:10.1371/journal.pone.0170650.

Cheung, H. O., X. Zhang, A. Ribeiro, R. Mo, S. Makino, V. Puviindran, K. K. Law, J. Briscoe, and C. C. Hui. 2009. "The kinesin protein Kif7 is a critical regulator of Gli transcription factors in mammalian hedgehog signaling." *Sci Signal* 2 (76):ra29. doi:10.1126/scisignal.2000405.

Dafinger, C., M. C. Liebau, S. M. Elsayed, Y. Hellenbroich, E. Boltshauser, G. C. Korenke, F. Fabretti, A. R. Janecke, I. Ebermann, G. Nurnberg, P. Nurnberg, H. Zentgraf, F. Koerber, K. Addicks, E. Elsobky, T. Benzing, B. Schermer, and H. J. Bolz. 2011. "Mutations in KIF7 link Joubert syndrome with Sonic Hedgehog signaling and microtubule dynamics." *J Clin Invest* 121 (7):2662–7. doi:10.1172/JCI43639.

Endoh-Yamagami, S., M. Evangelista, D. Wilson, X. Wen, J. W. Theunissen, K. Phamluong, M. Davis, S. J. Scales, M. J. Solloway, F. J. de Sauvage, and A. S. Peterson. 2009. "The mammalian Cos2 homolog Kif7 plays an essential role in modulating Hh signal transduction during development." *Curr Biol* 19 (15):1320–6. doi:10.1016/j.cub.2009.06.046.

Gao, J., N. Sai, C. Wang, X. Sheng, Q. Shao, C. Zhou, Y. Shi, S. Sun, X. Qu, and C. Zhu. 2011. "Overexpression of chromokinesin KIF4 inhibits proliferation of human gastric carcinoma cells both in vitro and in vivo." *Tumour Biol* 32 (1):53–61. doi:10.1007/s13277-010-0090-0.

Ghiretti, A. E., E. Thies, M. K. Tokito, T. Lin, E. M. Ostap, M. Kneussel, and E. L. F. Holzbaur. 2016. "Activity-dependent regulation of distinct transport and cytoskeletal remodeling functions of the dendritic kinesin KIF21B." *Neuron* 92 (4):857–72. doi:10.1016/j.neuron.2016.10.003.

Ha, M. J., J. Yoon, E. Moon, Y. M. Lee, H. J. Kim, and W. Kim. 2000. "Assignment of the kinesin family member 4 genes (KIF4A and KIF4B) to human chromosome bands Xq13.1 and 5q33.1 by in situ hybridization." *Cytogenet Cell Genet* 88 (1–2):41–2. doi:10.1159/000015482.

He, M., R. Subramanian, F. Bangs, T. Omelchenko, K. F. Liem, Jr., T. M. Kapoor, and K. V. Anderson. 2014. "The kinesin-4 protein Kif7 regulates mammalian Hedgehog signalling by organizing the cilium tip compartment." *Nat Cell Biol* 16 (7):663–72. doi:10.1038/ncb2988.

Heath, C. M., and S. M. Wignall. 2019. "Chromokinesin Kif4 promotes proper anaphase in mouse oocyte meiosis." *Mol Biol Cell* 30 (14):1691–704. doi:10.1091/mbc.E18-10-0666.

Heintz, T. G., J. P. Heller, R. Zhao, A. Caceres, R. Eva, and J. W. Fawcett. 2014. "Kinesin KIF4A transports integrin beta1 in developing axons of cortical neurons." *Mol Cell Neurosci* 63:60–71. doi:10.1016/j.mcn.2014.09.003.

Hou, G., C. Dong, Z. Dong, G. Liu, H. Xu, L. Chen, L. Liu, H. Wang, and W. Zhou. 2017. "Upregulate KIF4A enhances proliferation, invasion of hepatocellular carcinoma and indicates poor prognosis across human cancer types." *Sci Rep* 7 (1):4148. doi:10.1038/s41598-017-04176-9.

Hu, C. K., M. Coughlin, C. M. Field, and T. J. Mitchison. 2011. "KIF4 regulates midzone length during cytokinesis." *Curr Biol* 21 (10):815–24. doi:10.1016/j.cub.2011.04.019.

Jiang, S., N. Mani, E. M. Wilson-Kubalek, P. I. Ku, R. A. Milligan, and R. Subramanian. 2019. "Interplay between the kinesin and tubulin mechanochemical cycles underlies microtubule tip tracking by the non-motile ciliary kinesin Kif7." *Dev Cell* 49 (5):711–30 e8. doi:10.1016/j.devcel.2019.04.001.

Kim, W., Y. Tang, Y. Okada, T. A. Torrey, S. K. Chattopadhyay, M. Pfleiderer, F. G. Falkner, F. Dorner, W. Choi, N. Hirokawa, and H. C. Morse, 3rd. 1998. "Binding of murine leukemia virus Gag polyproteins to KIF4, a microtubule-based motor protein." *J Virol* 72 (8):6898–901.

Klejnot, M., and F. Kozielski. 2012. "Structural insights into human Kif7, a kinesin involved in Hedgehog signalling." *Acta Crystallogr D Biol Crystallogr* 68 (Pt 2):154–9. doi:10.1107/S0907444911053042.

Kurasawa, Y., W. C. Earnshaw, Y. Mochizuki, N. Dohmae, and K. Todokoro. 2004. "Essential roles of KIF4 and its binding partner PRC1 in organized central spindle midzone formation." *EMBO J* 23 (16):3237–48. doi:10.1038/sj.emboj.7600347.

Kwon, M., S. Morales-Mulia, I. Brust-Mascher, G. C. Rogers, D. J. Sharp, and J. M. Scholey. 2004. "The chromokinesin, KLP3A, dives mitotic spindle pole separation during prometaphase and anaphase and facilitates chromatid motility." *Mol Biol Cell* 15 (1):219–33. doi:10.1091/mbc.e03-07-0489.

Lee, Y. M., and W. Kim. 2004. "Kinesin superfamily protein member 4 (KIF4) is localized to midzone and midbody in dividing cells." *Exp Mol Med* 36 (1):93–7. doi:10.1038/emm.2004.13.

Luan, Y. J., S. H. Liu, Y. G. Sun, X. Qu, F. C. Wei, Y. Xu, and P. S. Yang. 2019. "Whole genome expression microarray reveals novel roles for Kif4 in monocyte/macrophage cells." *Eur Rev Med Pharmacol Sci* 23 (16):7016–23. doi:10.26355/eurrev_201908_18743.

Martinez, N. W., X. Xue, R. G. Berro, G. Kreitzer, and M. D. Resh. 2008. "Kinesin KIF4 regulates intracellular trafficking and stability of the human immunodeficiency virus type 1 Gag polyprotein." *J Virol* 82 (20):9937–50. doi:10.1128/JVI.00819-08.

Mazumdar, M., and T. Misteli. 2005. "Chromokinesins: multitalented players in mitosis." *Trends Cell Biol* 15 (7):349–55. doi:10.1016/j.tcb.2005.05.006.

Mazumdar, M., S. Sundareshan, and T. Misteli. 2004. "Human chromokinesin KIF4A functions in chromosome condensation and segregation." *J Cell Biol* 166 (5):613–20. doi:10.1083/jcb.200401142.

Mazumdar, M., M. H. Sung, and T. Misteli. 2011. "Chromatin maintenance by a molecular motor protein." *Nucleus* 2 (6):591–600. doi:10.4161/nucl.2.6.18044.

Midorikawa, R., Y. Takei, and N. Hirokawa. 2006. "KIF4 motor regulates activity-dependent neuronal survival by suppressing PARP-1 enzymatic activity." *Cell* 125 (2):371–83. doi:10.1016/j.cell.2006.02.039.

Mitchison, T. J., P. Nguyen, M. Coughlin, and A. C. Groen. 2013. "Self-organization of stabilized microtubules by both spindle and midzone mechanisms in Xenopus egg cytosol." *Mol Biol Cell* 24 (10):1559–73. doi:10.1091/mbc.E12-12-0850.

Nunes Bastos, R., S. R. Gandhi, R. D. Baron, U. Gruneberg, E. A. Nigg, and F. A. Barr. 2013. "Aurora B suppresses microtubule dynamics and limits central spindle size by locally activating KIF4A." *J Cell Biol* 202 (4):605–21. doi:10.1083/jcb.201301094.

Oh, S., H. Hahn, T. A. Torrey, H. Shin, W. Choi, Y. M. Lee, H. C. Morse, and W. Kim. 2000. "Identification of the human homologue of mouse KIF4, a kinesin superfamily motor protein." *Biochim Biophys Acta* 1493 (1–2):219–24. doi:10.1016/s0167-4781(00)00151-2.

Peretti, D., L. Peris, S. Rosso, S. Quiroga, and A. Caceres. 2000. "Evidence for the involvement of KIF4 in the anterograde transport of L1-containing vesicles." *J Cell Biol* 149 (1):141–52. doi:10.1083/jcb.149.1.141.

Powers, J., D. J. Rose, A. Saunders, S. Dunkelbarger, S. Strome, and W. M. Saxton. 2004. "Loss of KLP-19 polar ejection force causes misorientation and missegregation of holocentric chromosomes." *J Cell Biol* 166 (7):991–1001. doi:10.1083/jcb.200403036.

Putoux, A., S. Nampoothiri, N. Laurent, V. Cormier-Daire, P. L. Beales, A. Schinzel, D. Bartholdi, C. Alby, S. Thomas, N. Elkhartoufi, A. Ichkou, J. Litzler, A. Munnich, F. Encha-Razavi, R. Kannan, L. Faivre, N. Boddaert, A. Rauch, M. Vekemans, and T. Attie-Bitach. 2012. "Novel KIF7 mutations extend the phenotypic spectrum of acrocallosal syndrome." *J Med Genet* 49 (11):713–20. doi:10.1136/jmedgenet-2012-101016.

Putoux, A., S. Thomas, K. L. Coene, E. E. Davis, Y. Alanay, G. Ogur, E. Uz, D. Buzas, C. Gomes, S. Patrier, C. L. Bennett, N. Elkhartoufi, M. H. Frison, L. Rigonnot, N. Joye, S. Pruvost, G. E. Utine, K. Boduroglu, P. Nitschke, L. Fertitta, C. Thauvin-Robinet, A.

Munnich, V. Cormier-Daire, R. Hennekam, E. Colin, N. A. Akarsu, C. Bole-Feysot, N. Cagnard, A. Schmitt, N. Goudin, S. Lyonnet, F. Encha-Razavi, J. P. Siffroi, M. Winey, N. Katsanis, M. Gonzales, M. Vekemans, P. L. Beales, and T. Attie-Bitach. 2011. "KIF7 mutations cause fetal hydrolethalus and acrocallosal syndromes." *Nat Genet* 43 (6):601–6. doi:10.1038/ng.826.

Samejima, K., I. Samejima, P. Vagnarelli, H. Ogawa, G. Vargiu, D. A. Kelly, F. de Lima Alves, A. Kerr, L. C. Green, D. F. Hudson, S. Ohta, C. A. Cooke, C. J. Farr, J. Rappsilber, and W. C. Earnshaw. 2012. "Mitotic chromosomes are compacted laterally by KIF4 and condensin and axially by topoisomerase IIalpha." *J Cell Biol* 199 (5):755–70. doi:10.1083/jcb.201202155.

Sekine, Y., Y. Okada, Y. Noda, S. Kondo, H. Aizawa, R. Takemura, and N. Hirokawa. 1994. "A novel microtubule-based motor protein (KIF4) for organelle transports, whose expression is regulated developmentally." *J Cell Biol* 127 (1):187–201. doi:10.1083/jcb.127.1.187.

Sheng, L., S. L. Hao, W. X. Yang, and Y. Sun. 2018. "The multiple functions of kinesin-4 family motor protein KIF4 and its clinical potential." *Gene* 678:90–99. doi:10.1016/j.gene.2018.08.005.

Subramanian, R., S. C. Ti, L. Tan, S. A. Darst, and T. M. Kapoor. 2013. "Marking and measuring single microtubules by PRC1 and kinesin-4." *Cell* 154 (2):377–90. doi:10.1016/j.cell.2013.06.021.

Takahashi, M., T. Wakai, and T. Hirota. 2016. "Condensin I-mediated mitotic chromosome assembly requires association with chromokinesin KIF4A." *Genes Dev* 30 (17):1931–6. doi:10.1101/gad.282855.116.

van der Vaart, B., W. E. van Riel, H. Doodhi, J. T. Kevenaar, E. A. Katrukha, L. Gumy, B. P. Bouchet, I. Grigoriev, S. A. Spangler, K. L. Yu, P. S. Wulf, J. Wu, G. Lansbergen, E. Y. van Battum, R. J. Pasterkamp, Y. Mimori-Kiyosue, N. Demmers, N. Olieric, I. V. Maly, C. C. Hoogenraad, and A. Akhmanova. 2013. "CFEOM1-associated kinesin KIF21A is a cortical microtubule growth inhibitor." *Dev Cell* 27 (2):145–60. doi:10.1016/j.devcel.2013.09.010.

van Riel, W. E., A. Rai, S. Bianchi, E. A. Katrukha, Q. Liu, A. J. Heck, C. C. Hoogenraad, M. O. Steinmetz, L. C. Kapitein, and A. Akhmanova. 2017. "Kinesin-4 KIF21B is a potent microtubule pausing factor." *Elife* 6. doi:10.7554/eLife.24746.

Vernos, I., J. Raats, T. Hirano, J. Heasman, E. Karsenti, and C. Wylie. 1995. "Xklp1, a chromosomal Xenopus kinesin-like protein essential for spindle organization and chromosome positioning." *Cell* 81 (1):117–27. doi:10.1016/0092-8674(95)90376-3.

Vogel, P., R. W. Read, G. M. Hansen, B. J. Payne, D. Small, A. T. Sands, and B. P. Zambrowicz. 2012. "Congenital hydrocephalus in genetically engineered mice." *Vet Pathol* 49 (1):166–81. doi:10.1177/0300985811415708.

Wan, Q., Y. Shen, H. Zhao, B. Wang, L. Zhao, Y. Zhang, X. Bu, M. Wan, and C. Shen. 2019. "Impaired DNA double-strand breaks repair by kinesin family member 4A inhibition renders human H1299 non-small-cell lung cancer cells sensitive to cisplatin." *J Cell Physiol* 234 (7):10360–71. doi:10.1002/jcp.27703.

Wandke, C., M. Barisic, R. Sigl, V. Rauch, F. Wolf, A. C. Amaro, C. H. Tan, A. J. Pereira, U. Kutay, H. Maiato, P. Meraldi, and S. Geley. 2012. "Human chromokinesins promote chromosome congression and spindle microtubule dynamics during mitosis." *J Cell Biol* 198 (5):847–63. doi:10.1083/jcb.201110060.

Wang, H., C. Lu, Q. Li, J. Xie, T. Chen, Y. Tan, C. Wu, and J. Jiang. 2014. "The role of Kif4A in doxorubicin-induced apoptosis in breast cancer cells." *Mol Cells* 37 (11):812–8. doi:10.14348/molcells.2014.0210.

Wignall, S. M., and A. M. Villeneuve. 2009. "Lateral microtubule bundles promote chromosome alignment during acentrosomal oocyte meiosis." *Nat Cell Biol* 11 (7):839–44. doi:10.1038/ncb1891.

Willemsen, M. H., W. Ba, W. M. Wissink-Lindhout, A. P. de Brouwer, S. A. Haas, M. Bienek, H. Hu, L. E. Vissers, H. van Bokhoven, V. Kalscheuer, N. Nadif Kasri, and T. Kleefstra. 2014. "Involvement of the kinesin family members KIF4A and KIF5C in intellectual disability and synaptic function." *J Med Genet* 51 (7):487–94. doi:10.1136/jmedgenet-2013-102182.

Williams, B. C., M. F. Riedy, E. V. Williams, M. Gatti, and M. L. Goldberg. 1995. "The Drosophila kinesin-like protein KLP3A is a midbody component required for central spindle assembly and initiation of cytokinesis." *J Cell Biol* 129 (3):709–23. doi:10.1083/jcb.129.3.709.

Wilson, C. W., C. T. Nguyen, M. H. Chen, J. H. Yang, R. Gacayan, J. Huang, J. N. Chen, and P. T. Chuang. 2009. "Fused has evolved divergent roles in vertebrate Hedgehog signalling and motile ciliogenesis." *Nature* 459 (7243):98–102. doi:10.1038/nature07883.

Wu, G., and P. L. Chen. 2008. "Structural requirements of chromokinesin Kif4A for its proper function in mitosis." *Biochem Biophys Res Commun* 372 (3):454–8. doi:10.1016/j.bbrc.2008.05.065.

Wu, G., L. Zhou, L. Khidr, X. E. Guo, W. Kim, Y. M. Lee, T. Krasieva, and P. L. Chen. 2008. "A novel role of the chromokinesin Kif4A in DNA damage response." *Cell Cycle* 7 (13):2013–20. doi:10.4161/cc.7.13.6130.

Xu, Y., Y. Luan, S. Liu, J. Sun, K. Wang, J. Cai, W. Jiang, P. Yang, F. Wei, and X. Qu. 2017. "Kif4 regulates the expression of VEGFR1 through the PI3K/Akt signaling pathway in RAW264.7 monocytes/macrophages." *Int J Mol Med* 39 (5):1285–90. doi:10.3892/ijmm.2017.2936.

Xue, D., P. Cheng, M. Han, X. Liu, L. Xue, C. Ye, K. Wang, and J. Huang. 2018. "An integrated bioinformatical analysis to evaluate the role of KIF4A as a prognostic biomarker for breast cancer." *Onco Targets Ther* 11:4755–68. doi:10.2147/OTT.S164730.

Yamada, K., C. Andrews, W. M. Chan, C. A. McKeown, A. Magli, T. de Berardinis, A. Loewenstein, M. Lazar, M. O'Keefe, R. Letson, A. London, M. Ruttum, N. Matsumoto, N. Saito, L. Morris, M. Del Monte, R. H. Johnson, E. Uyama, W. A. Houtman, B. de Vries, T. J. Carlow, B. L. Hart, N. Krawiecki, J. Shoffner, M. C. Vogel, J. Katowitz, S. M. Goldstein, A. V. Levin, E. C. Sener, B. T. Ozturk, A. N. Akarsu, M. C. Brodsky, F. Hanisch, R. P. Cruse, A. A. Zubcov, R. M. Robb, P. Roggenkaemper, I. Gottlob, L. Kowal, R. Battu, E. I. Traboulsi, P. Franceschini, A. Newlin, J. L. Demer, and E. C. Engle. 2003. "Heterozygous mutations of the kinesin KIF21A in congenital fibrosis of the extraocular muscles type 1 (CFEOM1)." *Nat Genet* 35 (4):318–21. doi:10.1038/ng1261.

Yue, Y., T. L. Blasius, S. Zhang, S. Jariwala, B. Walker, B. J. Grant, J. C. Cochran, and K. J. Verhey. 2018. "Altered chemomechanical coupling causes impaired motility of the kinesin-4 motors KIF27 and KIF7." *J Cell Biol* 217 (4):1319–34. doi:10.1083/jcb.201708179.

Zhu, C., J. Zhao, M. Bibikova, J. D. Leverson, E. Bossy-Wetzel, J. B. Fan, R. T. Abraham, and W. Jiang. 2005. "Functional analysis of human microtubule-based motor proteins, the kinesins and dyneins, in mitosis/cytokinesis using RNA interference." *Mol Biol Cell* 16 (7):3187–99. doi:10.1091/mbc.e05-02-0167.

6 The Kinesin-5 Family
Transporters and Creators

*Mary Popov, Alina Goldstein-Levitin
and Larisa Gheber*

CONTENTS

The Kinesin-5 family are homotetrameric, typically plus-end-directed motors with the ability to slide anti-parallel microtubules and to alter microtubule dynamics. Yeast Kinesin-5s display the fascinating ability to switch direction of motility.

6.1 EXAMPLE FAMILY MEMBERS

Mammalian: HsEg5, Kif11
Xenopus laevis: Eg5
Drosophila melanogaster: Klp61F
Aspergillus nidulans: BimC
Schizosaccharomyces pombe: Cut7
Saccharomyces cerevisiae: Cin8, Kip1

6.2 STRUCTURAL INFORMATION

Kinesin-5 motors are unique in that they act as homotetramers, with pairs of catalytic motor domains located on opposite sides of a 60-nm-long rod-like mini-filament (Acar et al., 2013; Gordon and Roof, 1999; Kashina et al., 1996; Scholey et al., 2014) (Figure 6.1A). The motor domain is located N-terminally in the primary sequence and contains Kinesin-5-specific regions. The motor domain is followed by a flexible 14- to 18-amino acid neck linker, then a stalk region, containing stretches of coiled-coil responsible for multimerisation and a C-terminal tail.

6.2.1 N-TERMINAL NON-MOTOR EXTENSION

The region from the N-terminal to the motor domain in Kinesin-5s is considerably longer than in the Kinesin-1 family (Goulet and Moores, 2013; Singh et al., 2018). Cryogenic electron microscopy and kinetic experiments indicate that the longer non-motor N-terminal region of the *Xenopus* Kinesin-5, Eg5, docks onto the motor domain in several nucleotide-states (Goulet et al., 2012; 2014). Bidirectional Kinesin-5s have longer and more divergent N-terminal extensions, as compared with Kinesin-1 and plus-end-directed Kinesin-5 motors (Singh et al., 2018).

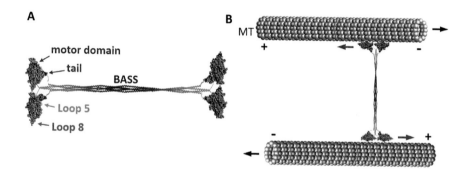

FIGURE 6.1 Schematic representation of a full-length Kinesin-5 tetramer and its arrangement when crosslinking spindle microtubules. Adapted from Singh et al. (2018). (A) The motor and tail domains are found at either end of the bipolar structure, connected through the central stalk that consists mainly of a coiled-coil structure and includes the bipolar-assembly (BASS) domain. The models of Cin8 motor and tail domains were constructed by homology modelling using the Swiss Model server (Arnold et al., 2006) and depicted using UCSF Chimera (Pettersen et al., 2004). The motor domain is superimposed on the cryogenic electron microscopy structure of a *S. pombe* Cut7 motor domain-decorated microtubule in the adenylyl-imidodiphosphate (AMP-PNP)-bound state (PDB: 5M5I) (Britto et al., 2016). (B) Schematic representation of the interaction of Kinesin-5 with two anti-parallel microtubules. Blue arrows indicate the direction of Kinesin-5 movement and black arrows represent the direction of microtubule movement during anti-parallel sliding.

6.2.2 LOOP 5

Loop 5 of the Kinesin-5 motor domain is long compared with other kinesin families, typically consisting of 18 residues (Behnke-Parks et al., 2011). The conformation of Loop 5 changes between "open" and "closed" conformations, affecting the ATPase cycle and the microtubule affinity of the Kinesin-5 motor domain (Behnke-Parks et al., 2011; Cochran and Gilbert, 2005; Larson et al., 2010; Waitzman et al., 2011). Coordinated conformational changes in Loop 5, the nucleotide-binding site and the neck linker during the ATPase cycle have been observed in solution (Larson et al., 2010; Maliga et al., 2006) and are abolished upon deletion of a portion of Loop 5 (Larson et al., 2010). Specific point mutations affecting proline residues within Loop 5 decreased both microtubule and nucleotide affinity and slowed Loop 5-dependent structural rearrangements that control neck linker docking (Behnke-Parks et al., 2011). Deletion of Loop 5 decreases the rate of microtubule-stimulated ADP release by Kinesin-5 monomers and dimers (Waitzman et al., 2011). Loop 5 also plays a role in synchronising the motor domains of Kinesin-5 dimers, thus enabling initiation of stepping from a two-head microtubule-bound state (Krzysiak and Gilbert, 2006; Waitzman et al., 2011).

Loop 5 is the binding site for small-molecule inhibitors specific to vertebrate Kinesin-5s (Asraf et al., 2015; Cochran et al., 2006; Kwok et al., 2006; Lakamper et al., 2010; Leizerman et al., 2004; Mayer et al., 1999; Yan et al., 2004). Available data suggest a mechanism whereby allosteric inhibitors bind to a specific conformation of Loop 5 and prevent any subsequent rearrangement of the motor domain required for the catalytic cycle (Brier et al., 2006; Kim et al., 2010; Maliga and Mitchison, 2006; Marshall et al., 2009).

6.2.3 NECK LINKER

A flexible 14- to 18-amino acid long neck linker immediately follows the motor domain and undergoes ATP- and microtubule-dependent docking onto the motor domain. The Kinesin-5 neck linker is longer than that of Kinesin-1, a trait that is suggested to contribute to the relatively low processivity of Kinesin-5 motors (Duselder et al., 2012; Shastry and Hancock, 2011). It was recently reported that the neck linker of Eg5 assumes different conformations, as compared with Kinesin-1, in some nucleotide-bound states (Muretta et al., 2015).

6.2.4 STALK REGION

The Kinesin-5 stalk contains four regions of heptad repeat sequences that form an α-helical coiled-coil, responsible for multimerisation. Deletion studies on the *S. cerevisiae* Kinesin-5, Cin8, and comparisons with other Kinesin-5 proteins suggest that the coiled-coil region located immediately after the neck linker (see below) is essential for self-interaction and sufficient for Cin8 dimerisation (Hildebrandt et al., 2006). The central bipolar assembly domain (BASS) spans ~200 residues in the central part of the stalk (Figure 6.1A) and is essential for Kinesin-5 activity and cell viability (Hildebrandt et al., 2006; Tao et al., 2006). The crystal structure of the *D. melanogaster* Kinesin-5 Klp61F BASS domain reveals that it consists of two anti-parallel coiled-coils, stabilised by alternating hydrophobic and ionic four-helical interfaces (Scholey et al., 2014). The helices emerge from the central part of the domain towards the N-terminal, where they bend, swap partners and form parallel coiled-coils offset by 90°. Based on this structure, it has been proposed that the central BASS domain plays a role in transmitting forces between motors situated at the opposite ends of the molecule (Fakhri and Schmidt, 2014; Scholey et al., 2014).

6.2.5 The C-terminal Tail Domain

The C-terminal tail domain is shown to be essential for microtubule crosslinking (Figure 6.1B). The Kinesin-5 tail contains an important Cdk1 (p34/Cdc2) kinase phosphorylation site (Figure 6.1A). In higher eukaryotes, this site is located within a conserved "BimC box" that is reportedly phosphorylated during mitosis (Blangy et al., 1995).

6.3 FUNCTIONAL PROPERTIES

6.3.1 Velocity, Processivity and Anti-Parallel Microtubule Sliding

The microtubule-stimulated ATPase rate of monomeric Kinesin-5 is slower than that of Kinesin-1 (~7/s vs. ~50/s) (Cochran et al., 2004; Cross, 2004; Rosenfeld et al., 2005), with phosphate release being the rate-limiting step (Cochran et al., 2006). Since the motor domain is located at the N-terminus, Kinesin-5s were initially thought to be exclusively plus-end-directed. In multi-motor microtubule gliding assays, monomeric and dimeric human, mouse and *X. laevis* Kinesin-5s were shown to move microtubules with the minus-ends leading, consistent with a plus-end directionality (Duselder et al., 2012; Kaseda et al., 2008; Sadakane et al., 2018; Yajima et al., 2008). The velocity of microtubule gliding was similar for monomers and dimers, suggesting that to translocate microtubules, coordination between the two motor domains within a dimer is not essential (Kaseda et al., 2008; Yajima et al., 2008). Plus-end-directed motility of full-length Kinesin-5 from human, *D. melanogaster*, *X. laevis* and *S. cerevisiae* was also demonstrated (Cole et al., 1994; Duselder et al., 2012; Fridman et al., 2013; Gheber et al., 1999; Kapitein et al., 2008; Kapitein et al., 2005; Roostalu et al., 2011; Sawin and Mitchison, 1995). However, the velocity of Kinesin-5-mediated microtubule gliding was considerably slower (10–70 nm/s) (Cochran, 2015; Wojcik et al., 2013) than that of Kinesin-1 motors (~500–1000 nm/s), indicating differences in the mechanochemical cycle (Waitzman and Rice, 2014). In single-molecule motility assays, Kinesin-5 from vertebrates displayed slow, plus-end-directed motility (Chen and Hancock, 2015; Duselder et al., 2012; Kapitein et al., 2008; Kwok et al., 2006; Shastry and Hancock, 2011).

The processivity of Kinesin-5s is relatively low. A truncated dimeric version of human Kinesin-5 takes on average approximately eight steps before detaching (Valentine et al., 2006). The processivity of full-length Kinesin-5 is higher than that of truncated dimeric constructs, but still lower than that of Kinesin-1 motors (Kwok et al., 2006). It is suggested that the longer neck-linker region is associated with the reduced processivity of Kinesin-5 motors (Duselder et al., 2012; Shastry and Hancock, 2011).

Crosslinking and sliding of anti-parallel microtubules has been demonstrated for Kinesin-5s from yeast to vertebrates (Gerson-Gurwitz et al., 2011; Gheber et al., 1999; Kapitein et al., 2008; Roostalu et al., 2011; van den Wildenberg et al., 2008). On a single microtubule, *X. laevis* Eg5 exhibits diffusive non-directed motility, which may be attributed to an additional microtubule-binding site in the tail (Weinger et al., 2011). This diffusive mode switches to plus-end-directed processive motility upon crosslinking of a second microtubule to achieve anti-parallel sliding (Kapitein et al., 2008), with force production during anti-parallel sliding being correlated with the length of the zone of overlap (Shimamoto et al., 2015).

Several studies have addressed the functionality of Kinesin-5 motors in the presence of other motors. *X. laevis* Eg5 was shown to antagonise and slow microtubule motility driven by the fast Kinesin-1 in gliding assays (Crevel et al., 2004), while the *D. melanogaster* Kinesin-5, Klp61F, was shown to antagonise the minus-end-directed Kinesin-14 Ncd (Tao et al., 2006).

6.3.2 BIDIRECTIONAL MOTILITY OF FUNGAL KINESIN-5

The *S. cerevisiae* Kinesin-5, Cin8, displays minus-end-directed motility as a single molecule under high ionic strength conditions, but switches directionality in multi-motor gliding and sliding assays and under low ionic strength conditions (Gerson-Gurwitz et al., 2011; Roostalu et al., 2011). The *S. cerevisiae* Kinesin-5, Kip1, and *S. pombe* Kinesin-5, Cut7, are also reported to exhibit switchable directionality (Edamatsu, 2014; Fridman et al., 2013). Replacement of the 99-amino acid insert in Loop 8 of the motor domain with the short Loop 8 of Kip1 results in bias towards minus-end-directed stepping (Gerson-Gurwitz et al., 2011). The switch to plus-end motility may be induced by coupling through crosslinked microtubules; this model is based on the observation that the switch is dependent on microtubule length (Roostalu et al., 2011). A second model suggests that the directionality of motility depends on the local density of motors (Britto et al., 2016). For Cut7, minus-end- directed stepping is selectively inhibited under crowded conditions, whereas plus-end-directed stepping is not. Cin8 is seen to accumulate into clusters under high ionic strength conditions, which slows minus-end-directed motility and induces a switch to plus-end-directed motility (Shapira et al., 2017). The mechanisms that regulate directional switching of these members of the Kinesin-5 family remain to be fully elucidated.

6.3.3 MICROTUBULE POLYMERISATION

Members of the Kinesin-5 family have been observed to alter microtubule growth dynamics. *S. cerevisiae* Kip1 tracks the plus-end of growing and shrinking microtubules in cells (Fridman et al., 2013) and Cin8 accumulates at and tracks the minus-ends of dynamic microtubules *in vitro* (Shapira et al., 2017). A dimeric chimera, comprising the *X. laevis* Eg5 motor domain and a Kinesin-1 stalk, accumulates at the microtubule plus-end and promotes polymerisation (Chen and Hancock,

2015). However, deletion of *S. cerevisiae* Cin8 or inactivation of the *D. melanogaster* Klp61F results in longer and more stable microtubules (Fridman et al., 2009; Gardner et al., 2008; Tubman et al., 2018), indicating that they have microtubule-destabilising activity.

6.4 PHYSIOLOGICAL ROLES

A table listing the currently identified physiological roles of Kinesin-5s is available in the online material (Table 6.1).

6.4.1 BIPOLAR SPINDLE ASSEMBLY, MAINTENANCE AND ELONGATION

Numerous studies have indicated that loss of Kinesin-5 function leads to failure of mitosis due to a lack of spindle pole separation prior to spindle assembly (Figure 6.2), or as a result of spindle collapse after establishment of the bipolar spindle (Bannigan et al., 2007; Blangy et al., 1995; Castillo and Justice, 2007; Heck et al., 1993; Hoyt et al., 1992; Kapoor et al., 2000; Rusan et al., 2002; Sawin et al., 1992). However, it remains to be fully established whether or not anti-parallel microtubule sliding is essential for spindle assembly, or if microtubule crosslinking by Kinesin-5 motors is sufficient (Crasta et al., 2006). To date, only the nematode *Caenorhabditis elegans* and the slime mould *Dictyostelium discoideum* have been reported to be able to form a functional bipolar spindle in the absence of Kinesin-5 (Bishop et al., 2005; Tikhonenko et al., 2008).

The role of Kinesin-5 in spindle elongation remains controversial. An anaphase-facilitating function (Figure 6.2D) was demonstrated in *S. cerevisiae* (Gerson-Gurwitz et al., 2009; Movshovich et al., 2008; Saunders et al., 1995; Straight et al.,

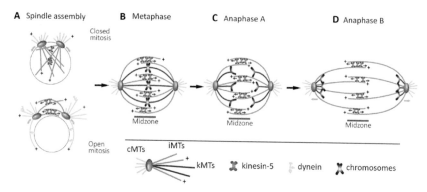

FIGURE 6.2 Schematic representation of the major roles of Kinesin-5s in mitotic spindle dynamics. Adapted from Singh et al. (2018). (A) Spindle pole separation during spindle assembly in closed (top) and open (bottom) mitosis. The direction of movement of Kinesin-5 and the spindle poles are indicated by the blue and brown arrows, respectively; cMTs, iMTs and kMTs designate cytoplasmic, interpolar and kinetochore microtubules (MTs), respectively. (B) Chromosome congression in metaphase. Kinesin-5 motors crosslink anti-parallel iMTs at the midzone and stabilise the spindle. (C) Anaphase A: sister chromatids are pulled to the opposite spindle poles. (D) Anaphase B: spindle elongation is marked by separation of the two spindle poles. Spindle elongation is mediated by cortical force generators, such as Kinesin-5-mediated forces produced by sliding anti-parallel iMTs apart at the midzone.

1998) and insect cells (Sharp et al., 1999b), and the *S. cerevisiae* Kinesin-5s, Cin8 and Kip1, were shown to be partially destabilised in order to prevent anaphase onset in cells with DNA damage (Zhang et al., 2009). It is suggested that Kinesin-5s slow spindle elongation by applying "brakes" via the crosslinking of spindle microtubules (Collins et al., 2014; Rozelle et al., 2011; Saunders et al., 2007; Tikhonenko et al., 2008).

Biophysical models describing the role of Kinesin-5 motors in maintaining spindle bipolarity have been proposed, based on a force-balanced model suggested for *S. cerevisiae* (Hoyt et al., 1993; Saunders et al., 1997). In this model, bipolarity of the spindle is maintained by a balance of inward- and outward-directed forces exerted by the minus-end-directed Kinesin-14, Kar3, and the Kinesin-5s, Cin8 and Kip1, which are plus-end-directed between anti-parallel microtubules (Fridman et al., 2013; Gerson-Gurwitz et al., 2011; Roostalu et al., 2011). Accordingly, in *S. pombe*, bipolar spindles could not form when Kinesin-5 was deleted but could be assembled when Kinesin-5 and Kinesin-14 were simultaneously deleted (Rincon et al., 2017).

In higher eukaryotic cells, Kinesin-5s are suggested to affect the poleward turnover of tubulin (poleward flux) in both kinetochore and interpolar microtubules (Miyamoto et al., 2004; Rogers et al., 2005), contributing to chromosome congression and separation, respectively (Brust-Mascher et al., 2004, 2009; Sharp et al., 1999a). In *S. cerevisiae*, Kinesin-5s are shown to affect microtubule dynamics (Fridman et al., 2009; Fridman et al., 2013; Gardner et al., 2008) and have been shown to bind to kinetochores, to focus kinetochore clusters and to limit the length of kinetochore microtubules (De Wulf et al., 2003; Gardner et al., 2008; Suzuki et al., 2018; Tytell and Sorger, 2006; Wargacki et al., 2010).

6.4.2 FUNCTIONS AT THE SPINDLE POLES

Several pieces of evidence suggest that localisation near the spindle poles is important for Kinesin-5 function. First, in a number of organisms, Kinesin-5 motors were found to be enriched near centrosomes or spindle pole bodies (Cheerambathur et al., 2008; Gable et al., 2012; Miki et al., 2014; Sawin et al., 1992; Sawin and Mitchison, 1995; Uteng et al., 2008). Second, in higher eukaryotes undergoing open mitosis, Kinesin-5s are actively transported towards the poles by the dynein-dynactin complex (Gable et al., 2012; Kapoor and Mitchison, 2001; Uteng et al., 2008). The Ran-regulated spindle pole-localising factor TPX2 (Wittmann et al., 2000) was found to recruit Kinesin-5 to the poles in *Xenopus* spindles (Helmke and Heald, 2014), possibly via coupling to the dynein complex (Ma et al., 2011). Kinesin-5s also co-localise with spindle poles prior to spindle assembly in yeast (Hagan and Yanagida, 1990; Shapira et al., 2017). Bidirectionality may be the evolutionary adaptation that allows these motors to localise to spindle poles prior to spindle assembly, without the need for the minus-end-directed dynein (Mann and Wadsworth, 2019; Shapira et al., 2017; Singh et al., 2018). At the spindle poles, Kinesin-5s can capture anti-parallel microtubules emanating from the opposite poles and crosslink them (Crasta et al., 2006; Gheber et al., 1999) or mediate their anti-parallel sliding, thus promoting spindle assembly (Shapira et al., 2017).

6.4.3 REGULATION BY POST-TRANSLATIONAL MODIFICATION

Kinesin-5s contain multiple phosphorylation sites located within and outside the motor domain (Figure 6.3). In several Kinesin-5s, a conserved sequence termed the "BimC box" is located in the C-terminal tail and contains a conserved phosphorylation site for cyclin-dependent kinase 1 (Cdk1) (Figure 6.2A). In human Eg5, phosphorylation at this site regulates spindle targeting, association and localisation (Blangy et al., 1995). *X. laevis* mutants phosphodeficient in the Eg5 BimC box show disrupted kinase localisation to the mitotic spindle (Sawin and Mitchison, 1995), whereas in spindles assembled from *Xenopus* egg extract, phosphorylation by Cdk1 increased the binding of Eg5 to microtubules (Cahu et al., 2008). In higher plants, microtubule stabilisation and anti-parallel microtubule sliding by Kinesin-5 is Cdk1-dependent (Barroso et al., 2000; Hemerly et al., 1993; Hemsley et al., 2001; Lee et al., 2001). In *D. melanogaster* embryonic mitotic spindles, Klp61F, phosphorylation within the BimC box, promotes its concentration in the spindle mid-zone (Sharp et al., 1999a). Klp61F is phosphorylated both within the BimC box and in the motor domain (Cheerambathur et al., 2008; Garcia et al., 2009), leading to dynamic localisation of Klp61F throughout the spindle and cross-bridging between both parallel and anti-parallel microtubules (Barton et al., 1995; Sharp et al., 1999a).

Cdk1 also phosphorylates Kinesin-5s in yeast, although the mechanism of phosphoregulation appears to be different from that in higher eukaryotes. In fission yeast, Cdk1 phosphorylation at the C-terminal region of Cut7 is not required for association with microtubules (Drummond and Hagan, 1998). Both *S. cerevisiae* Cin8 and Kip1

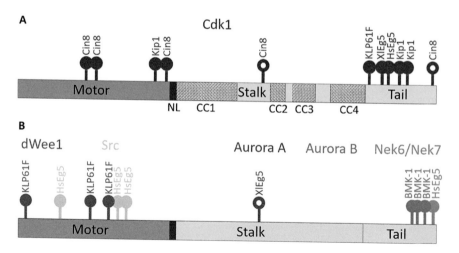

FIGURE 6.3 Location of experimentally identified phosphorylation sites in Kinesin-5s. Indicated by shades of grey from left to right: motor domain, neck linker (NL), the stalk containing putative coiled-coil (cc) regions and the tail domain. Locations of the phosphorylation sites are indicated by spheres. Filled spheres: identified function; empty spheres: unknown function. Phosphorylation sites are color coded according to the corresponding kinases. (A) Phosphorylation sites for Cdk1. (B) Phosphorylation sites for dWee1, Src, Aurora A, Aurora B and Nek6/Nek7 kinases.

lack the BimC box motif, although Cdk1 phosphorylation consensus sequences are present in their C-terminal tails (Chee and Haase, 2010). Phosphorylation of Kip1 in the motor domain by Cdk1 is required for spindle pole separation (Chee and Haase, 2010). Cin8 has been shown to be differentially phosphorylated during anaphase at three Cdk1 sites located in its motor domain (Avunie-Masala et al., 2011), with phosphorylation inducing Cin8 detachment from spindles and reducing the spindle elongation rate (Avunie-Masala et al., 2011), and with each of the three sites playing unique roles in Cin8 regulation (Goldstein et al., 2019; 2017). *In vitro* studies of Cin8 phospho-mutants suggest that phosphorylation of the three sites within the motor domain provides fine-tuning of motor activity (Shapira and Gheber, 2016).

The tail domain of human Eg5 contains a phosphorylation site recognized by the Nek6/Nek7 kinases. This site contributes to the accumulation of Eg5 at the spindle poles and is necessary for subsequent spindle pole separation (Bertran et al., 2011; Rapley et al., 2008). Phosphorylation of the Eg5 motor domain by Src kinase fine-tunes motor activity and promotes optimal spindle morphology (Bickel et al., 2017). The *D. melanogaster* Kinesin-5, Klp61F, is phosphorylated in the motor domain by the Wee1 tyrosine kinase, which may affect Loop 5 conformation and motor function (Garcia et al., 2009). Phosphorylation by the Aurora kinases has also been reported for some Kinesin-5s. *C. elegans* Kinesin-5 is phosphorylated in the tail domain by Aurora B, which regulates its spindle localisation (Bishop et al., 2005; Cahu et al., 2008). *X. laevis* Eg5 is phosphorylated by Aurora A in the stalk domain (Giet et al., 1999), although this was found to be non-essential for spindle formation (Cahu et al., 2008).

Human Eg5 is acetylated at a specific lysine residue (K146) in its motor domain. The acetylation-mimic mutant, K146Q, disrupts the formation of a salt bridge, converting Eg5 into a motor that is more resistant to dissociation from microtubules under load (Muretta et al., 2018).

6.4.4 Roles in Non-Dividing Cells

Several studies have revealed roles for Kinesin-5s in interphase, including regulation of the biogenesis and function of Ago1-complexes (Stoica et al., 2010), as well as Golgi organization and polypeptide synthesis (Bartoli et al., 2011; Whitehead and Rattner, 1998). The most extensively studied function of Kinesin-5 in interphase is in neuronal cells. Kinesin-5s are expressed in post-mitotic neurons (Lin et al., 2012), where they may play a role in the proper development of mammalian neuronal processes, including axon growth cone guidance, elongation and branching (Haque et al., 2004; Myers and Baas, 2007; Nadar et al., 2008) and modulation of neuronal growth and migration (Falnikar et al., 2011).

Several studies have revealed that Kinesin-5s play non-canonical roles in plant cells (Liu et al., 2018). In *Nicotiana tabacum*, Kinesin-5 was reported to be involved in separating anti-parallel microtubules in the phragmoplast, a microtubule-based structure formed during late cytokinesis (Asada et al., 1997). In *Arabidopsis thaliana*, disruption of Kinesin-5 activity leads to disorganisation of intracellular microtubules during interphase, as well as in spindle formation (Bannigan et al., 2007).

6.5   INVOLVEMENT IN DISEASE

The Kinesin-5, Eg5, is over-expressed in haematological malignancies and many solid tumours, such as breast, ovarian, bladder and pancreatic cancers (Exertier et al., 2013; Liu et al., 2010; Sun et al., 2015; Wang et al., 2017). In breast cancer patients, Eg5 over-expression is associated with poor prognosis and has been proposed as a potential prognostic biomarker and target for therapeutic agents in oral and breast cancers (Daigo et al., 2018; Pei et al., 2017). Eg5 possesses unique structural features that selectively predispose it to small-molecule inhibitors. In the Eg5 motor domain, Loop 5 is elongated, forming a druggable allosteric pocket (Turner et al., 2001). Moreover, Loop 5 is flexible (Muretta et al., 2013) and undergoes an "open" to "closed" structural transition that correlates with rearrangements at the active site during the ATPase cycle (Behnke-Parks et al., 2011; Cochran and Gilbert, 2005; Larson et al., 2010; Waitzman et al., 2011). Since the discovery of monastrol, a specific vertebrate Kinesin-5 inhibitor that binds to Loop 5 (Kapoor et al., 2000; Mayer et al., 1999), a variety of Eg5-specific inhibitors have been developed (Gartner et al., 2005; Huszar et al., 2009; Liu et al., 2013). These cause mitotic arrest, producing characteristic monoastral spindles (Asraf et al., 2015; Chen et al., 2017; Leizerman et al., 2004), and lead to apoptosis in proliferative tissues.

Several Kinesin-5 inhibitors are currently in clinical trials (Chan et al., 2012; Lee et al., 2008; Sarli and Giannis, 2008; Shah et al., 2017). However, more than 40 Phase I and II clinical trials assessing inhibitors of Eg5, starting with the first generation drug ispinesib (Lee et al., 2008) and followed by filanesib (Shah et al., 2017), have been suspended or discontinued. When used as a monotherapy, Eg5-targeting agents have been only moderately successful, instead causing adverse effects, such as neutropenia (Rath and Kozielski, 2012). The kinesin KIF15 can functionally replace Eg5, causing resistance to Eg5 inhibitors (Sturgill et al., 2016). To overcome this, a strategy involving a combination therapy, employing inhibitors to both kinesins, has been suggested (Milic et al., 2018).

ACKNOWLEDGEMENTS

This work was supported in part by Israel Science Foundation (ISF) grant 386/18 and United States–Israel Binational Science Foundation (BSF) grant BSF-2015851 awarded to L.G.

REFERENCES

Acar, S., D.B. Carlson, M.S. Budamagunta, V. Yarov-Yarovoy, J.J. Correia, M.R. Ninonuevo, W. Jia, L. Tao, J.A. Leary, J.C. Voss, J.E. Evans, and J.M. Scholey. 2013. The bipolar assembly domain of the mitotic motor kinesin-5. *Nat Commun.* 4:1343.

Arnold, K., L. Bordoli, J. Kopp, and T. Schwede. 2006. The SWISS-MODEL workspace: a web-based environment for protein structure homology modelling. *Bioinformatics.* 22:195–201.

Asada, T., R. Kuriyama, and H. Shibaoka. 1997. TKRP125, a kinesin-related protein involved in the centrosome- independent organization of the cytokinetic apparatus in tobacco BY-2 cells. *J Cell Sci.* 110:179–189.

Asraf, H., R. Avunie-Masala, M. Hershfinkel, and L. Gheber. 2015. Mitotic slippage and expression of survivin are linked to differential sensitivity of human cancer cell-lines to the Kinesin-5 inhibitor monastrol. *PLoS One.* 10:e0129255.

Avunie-Masala, R., N. Movshovich, Y. Nissenkorn, A. Gerson-Gurwitz, V. Fridman, M. Koivomagi, M. Loog, M.A. Hoyt, A. Zaritsky, and L. Gheber. 2011. Phospho-regulation of kinesin-5 during anaphase spindle elongation. *J Cell Sci.* 124:873–878.

Bannigan, A., W.R. Scheible, W. Lukowitz, C. Fagerstrom, P. Wadsworth, C. Somerville, and T.I. Baskin. 2007. A conserved role for kinesin-5 in plant mitosis. *J Cell Sci.* 120:2819–2827.

Barroso, C., J. Chan, V. Allan, J. Doonan, P. Hussey, and C. Lloyd. 2000. Two kinesin-related proteins associated with the cold-stable cytoskeleton of carrot cells: characterization of a novel kinesin, DcKRP120-2. *Plant J.* 24:859–868.

Bartoli, K.M., J. Jakovljevic, J.L. Woolford, Jr., and W.S. Saunders. 2011. Kinesin molecular motor Eg5 functions during polypeptide synthesis. *Mol Biol Cell.* 22:3420–3430.

Barton, N.R., A.J. Pereira, and L.S. Goldstein. 1995. Motor activity and mitotic spindle localization of the Drosophila kinesin-like protein KLP61F. *Mol Biol Cell.* 6:1563–1574.

Behnke-Parks, W.M., J. Vendome, B. Honig, Z. Maliga, C. Moores, and S.S. Rosenfeld. 2011. Loop L5 acts as a conformational latch in the mitotic kinesin Eg5. *J Biol Chem.* 286:5242–5253.

Bertran, M.T., S. Sdelci, L. Regue, J. Avruch, C. Caelles, and J. Roig. 2011. Nek9 is a Plk1-activated kinase that controls early centrosome separation through Nek6/7 and Eg5. *EMBO J.* 30:2634–2647.

Bickel, K.G., B.J. Mann, J.S. Waitzman, T.A. Poor, S.E. Rice, and P. Wadsworth. 2017. Src family kinase phosphorylation of the motor domain of the human kinesin-5, Eg5. *Cytoskeleton (Hoboken).* 74:317–330.

Bishop, J.D., Z. Han, and J.M. Schumacher. 2005. The Caenorhabditis elegans Aurora B kinase AIR-2 phosphorylates and is required for the localization of a BimC kinesin to meiotic and mitotic spindles. *Mol Biol Cell.* 16:742–756.

Blangy, A., H.A. Lane, P. d'Herin, M. Harper, M. Kress, and E.A. Nigg. 1995. Phosphorylation by p34cdc2 regulates spindle association of human Eg5, a kinesin-related motor essential for bipolar spindle formation in vivo. *Cell.* 83:1159–1169.

Brier, S., D. Lemaire, S. DeBonis, E. Forest, and F. Kozielski. 2006. Molecular dissection of the inhibitor binding pocket of mitotic kinesin Eg5 reveals mutants that confer resistance to antimitotic agents. *J Mol Biol.* 360:360–376.

Britto, M., A. Goulet, S. Rizvi, O. von Loeffelholz, C.A. Moores, and R.A. Cross. 2016. Schizosaccharomyces pombe kinesin-5 switches direction using a steric blocking mechanism. *Proc Natl Acad Sci U S A.* 113:E7483–E7489.

Brust-Mascher, I., G. Civelekoglu-Scholey, M. Kwon, A. Mogilner, and J.M. Scholey. 2004. Model for anaphase B: role of three mitotic motors in a switch from poleward flux to spindle elongation. *Proc Natl Acad Sci U S A.* 101:15938–15943.

Brust-Mascher, I., P. Sommi, D.K. Cheerambathur, and J.M. Scholey. 2009. Kinesin-5-dependent poleward flux and spindle length control in Drosophila embryo mitosis. *Mol Biol Cell.* 20:1749–1762.

Cahu, J., A. Olichon, C. Hentrich, H. Schek, J. Drinjakovic, C. Zhang, A. Doherty-Kirby, G. Lajoie, and T. Surrey. 2008. Phosphorylation by Cdk1 increases the binding of Eg5 to microtubules in vitro and in Xenopus egg extract spindles. *PLoS ONE.* 3:e3936.

Castillo, A., and M.J. Justice. 2007. The kinesin related motor protein, Eg5, is essential for maintenance of pre-implantation embryogenesis. *Biochem Biophys Res Commun.* 357:694–699.

Chan, K.S., C.G. Koh, and H.Y. Li. 2012. Mitosis-targeted anti-cancer therapies: where they stand. *Cell Death Dis.* 3:e411.

Chee, M.K., and S.B. Haase. 2010. B-cyclin/CDKs regulate mitotic spindle assembly by phosphorylating kinesins-5 in budding yeast. *PLoS Genet.* 6:e1000935.

Cheerambathur, D.K., I. Brust-Mascher, G. Civelekoglu-Scholey, and J.M. Scholey. 2008. Dynamic partitioning of mitotic kinesin-5 cross-linkers between microtubule-bound and freely diffusing states. *J Cell Biol.* 182:429–436.

Chen, G.Y., Y.J. Kang, A.S. Gayek, W. Youyen, E. Tuzel, R. Ohi, and W.O. Hancock. 2017. Eg5 inhibitors have contrasting effects on microtubule stability and metaphase spindle integrity. *ACS Chem Biol.* 12:1038–1046.

Chen, Y., and W.O. Hancock. 2015. Kinesin-5 is a microtubule polymerase. *Nat Commun.* 6:8160.

Cochran, J.C. 2015. Kinesin motor enzymology: chemistry, structure, and physics of nanoscale molecular machines. *Biophys Rev.* 7:269–299.

Cochran, J.C., and S.P. Gilbert. 2005. ATPase mechanism of Eg5 in the absence of microtubules: insight into microtubule activation and allosteric inhibition by monastrol. *Biochemistry.* 44:16633–16648.

Cochran, J.C., T.C. Krzysiak, and S.P. Gilbert. 2006. Pathway of ATP hydrolysis by monomeric kinesin Eg5. *Biochemistry.* 45:12334–12344.

Cochran, J.C., C.A. Sontag, Z. Maliga, T.M. Kapoor, J.J. Correia, and S.P. Gilbert. 2004. Mechanical analysis of the mitotic kinesin Eg5. *J Biol Chem.* 279:38861–38870.

Cole, D.G., W.M. Saxton, K.B. Sheehan, and J.M. Scholey. 1994. A "slow" homotetrameric kinesin-related motor protein purified from Drosophila embryos. *J Biol Chem.* 269:22913–22916.

Collins, E., B.J. Mann, and P. Wadsworth. 2014. Eg5 restricts anaphase B spindle elongation in mammalian cells. *Cytoskeleton (Hoboken).* 71:136–144.

Crasta, K., P. Huang, G. Morgan, M. Winey, and U. Surana. 2006. Cdk1 regulates centrosome separation by restraining proteolysis of microtubule-associated proteins. *EMBO J.* 25:2551–2563.

Crevel, I.M., M.C. Alonso, and R.A. Cross. 2004. Monastrol stabilises an attached low-friction mode of Eg5. *Curr Biol.* 14:R411–R412.

Cross, R.A. 2004. The kinetic mechanism of kinesin. *Trends Biochem Sci.* 29:301–309.

Daigo, K., A. Takano, P.M. Thang, Y. Yoshitake, M. Shinohara, I. Tohnai, Y. Murakami, J. Maegawa, and Y. Daigo. 2018. Characterization of KIF11 as a novel prognostic biomarker and therapeutic target for oral cancer. *Int J Oncol.* 52:155–165.

De Wulf, P., A.D. McAinsh, and P.K. Sorger. 2003. Hierarchical assembly of the budding yeast kinetochore from multiple subcomplexes. *Genes Dev.* 17:2902–2921.

Drummond, D.R., and I.M. Hagan. 1998. Mutations in the bimC box of Cut7 indicate divergence of regulation within the bimC family of kinesin related proteins. *J Cell Sci.* 111(Pt 7):853–865.

Duselder, A., C. Thiede, C.F. Schmidt, and S. Lakamper. 2012. Neck-linker length dependence of processive kinesin-5 motility. *J Mol Biol,* 423:159–168.

Edamatsu, M. 2014. Bidirectional motility of the fission yeast kinesin-5, Cut7. *Biochem Biophys Res Commun.* 446:231–234.

Exertier, P., S. Javerzat, B. Wang, M. Franco, J. Herbert, N. Platonova, M. Winandy, N. Pujol, O. Nivelles, S. Ormenese, V. Godard, J. Becker, R. Bicknell, R. Pineau, J. Wilting, A. Bikfalvi, and M. Hagedorn. 2013. Impaired angiogenesis and tumor development by inhibition of the mitotic kinesin Eg5. *Oncotarget.* 4:2302–2316.

Fakhri, N., and C.F. Schmidt. 2014. A surprising twist. *Elife.* 3:e02715.

Falnikar, A., S. Tole, and P.W. Baas. 2011. Kinesin-5, a mitotic microtubule-associated motor protein, modulates neuronal migration. *Mol Biol Cell.* 22:1561–1574.

Fridman, V., A. Gerson-Gurwitz, N. Movshovich, M. Kupiec, and L. Gheber. 2009. Midzone organization restricts interpolar microtubule plus-end dynamics during spindle elongation. *EMBO Rep.* 10:387–393.

Fridman, V., A. Gerson-Gurwitz, O. Shapira, N. Movshovich, S. Lakamper, C.F. Schmidt, and L. Gheber. 2013. Kinesin-5 Kip1 is a bi-directional motor that stabilizes microtubules and tracks their plus-ends in vivo. *J Cell Sci.* 126:4147–4159.

Gable, A., M. Qiu, J. Titus, S. Balchand, N.P. Ferenz, N. Ma, E.S. Collins, C. Fagerstrom, J.L. Ross, G. Yang, and P. Wadsworth. 2012. Dynamic reorganization of Eg5 in the mammalian spindle throughout mitosis requires dynein and TPX2. *Mol Biol Cell.* 23:1254–1266.

Garcia, K., J. Stumpff, T. Duncan, and T.T. Su. 2009. Tyrosines in the kinesin-5 head domain are necessary for phosphorylation by Wee1 and for mitotic spindle integrity. *Curr Biol.* 19:1670–1676.

Gardner, M.K., D.C. Bouck, L.V. Paliulis, J.B. Meehl, E.T. O'Toole, J. Haase, A. Soubry, A.P. Joglekar, M. Winey, E.D. Salmon, K. Bloom, and D.J. Odde. 2008. Chromosome congression by Kinesin-5 motor-mediated disassembly of longer kinetochore microtubules. *Cell.* 135:894–906.

Gartner, M., N. Sunder-Plassmann, J. Seiler, M. Utz, I. Vernos, T. Surrey, and A. Giannis. 2005. Development and biological evaluation of potent and specific inhibitors of mitotic Kinesin Eg5. *Chembiochem.* 6:1173–1177.

Gerson-Gurwitz, A., N. Movshovich, R. Avunie, V. Fridman, K. Moyal, B. Katz, M.A. Hoyt, and L. Gheber. 2009. Mid-anaphase arrest in S. cerevisiae cells eliminated for the function of Cin8 and dynein. *Cell Mol Life Sci.* 66:301–313.

Gerson-Gurwitz, A., C. Thiede, N. Movshovich, V. Fridman, M. Podolskaya, T. Danieli, S. Lakamper, D.R. Klopfenstein, C.F. Schmidt, and L. Gheber. 2011. Directionality of individual kinesin-5 Cin8 motors is modulated by loop 8, ionic strength and microtubule geometry. *EMBO J.* 30:4942–4954.

Gheber, L., S.C. Kuo, and M.A. Hoyt. 1999. Motile properties of the kinesin-related Cin8p spindle motor extracted from Saccharomyces cerevisiae cells. *J Biol Chem.* 274:9564–9572.

Giet, R., R. Uzbekov, F. Cubizolles, K. Le Guellec, and C. Prigent. 1999. The Xenopus laevis aurora-related protein kinase pEg2 associates with and phosphorylates the kinesin-related protein XlEg5. *J Biol Chem.* 274:15005–15013.

Goldstein, A., D. Goldman, E. Valk, M. Loog, L.J. Holt, and L. Gheber. 2019. Synthetic-evolution reveals narrow paths to regulation of the saccharomyces cerevisiae mitotic kinesin-5 Cin8. *Int J Bio Sci.* 15:1125–1138.

Goldstein, A., N. Siegler, D. Goldman, H. Judah, E. Valk, M. Koivomagi, M. Loog, and L. Gheber. 2017. Three Cdk1 sites in the kinesin-5 Cin8 catalytic domain coordinate motor localization and activity during anaphase. *Cell Mol Life Sci.* 74:3395–3412.

Gordon, D.M., and D.M. Roof. 1999. The kinesin-related protein Kip1p of Saccharomyces cerevisiae is bipolar. *J Biol Chem.* 274:28779–28786.

Goulet, A., W.M. Behnke-Parks, C.V. Sindelar, J. Major, S.S. Rosenfeld, and C.A. Moores. 2012. The structural basis of force generation by the mitotic motor kinesin-5. *J Biol Chem.* 287:44654–44666.

Goulet, A., J. Major, Y. Jun, S.P. Gross, S.S. Rosenfeld, and C.A. Moores. 2014. Comprehensive structural model of the mechanochemical cycle of a mitotic motor highlights molecular adaptations in the kinesin family. *Proc Natl Acad Sci U S A.* 111:1837–1842.

Goulet, A., and C. Moores. 2013. New insights into the mechanism of force generation by kinesin-5 molecular motors. *Int Rev Cell Mol Biol.* 304:419–466.

Hagan, I., and M. Yanagida. 1990. Novel potential mitotic motor protein encoded by the fission yeast cut7+ gene. *Nature.* 347:563–566.

Haque, S.A., T.P. Hasaka, A.D. Brooks, P.V. Lobanov, and P.W. Baas. 2004. Monastrol, a prototype anti-cancer drug that inhibits a mitotic kinesin, induces rapid bursts of axonal outgrowth from cultured postmitotic neurons. *Cell Motil Cytoskeleton.* 58:10–16.

Heck, M.M., A. Pereira, P. Pesavento, Y. Yannoni, A.C. Spradling, and L.S. Goldstein. 1993. The kinesin-like protein KLP61F is essential for mitosis in Drosophila. *J Cell Biol.* 123:665–679.

Helmke, K.J., and R. Heald. 2014. TPX2 levels modulate meiotic spindle size and architecture in Xenopus egg extracts. *J Cell Biol.* 206:385–393.

Hemerly, A.S., P. Ferreira, J. de Almeida Engler, M. Van Montagu, G. Engler, and D. Inze. 1993. cdc2a expression in Arabidopsis is linked with competence for cell division. *Plant Cell.* 5:1711–1723.

Hemsley, R., S. McCutcheon, J. Doonan, and C. Lloyd. 2001. P34(cdc2) kinase is associated with cortical microtubules from higher plant protoplasts. *FEBS Lett.* 508:157–161.

Hildebrandt, E.R., L. Gheber, T. Kingsbury, and M.A. Hoyt. 2006. Homotetrameric form of Cin8p, a Saccharomyces cerevisiae kinesin-5 motor, is essential for its in vivo function. *J Biol Chem.* 281:26004–26013.

Hoyt, M.A., L. He, K.K. Loo, and W.S. Saunders. 1992. Two Saccharomyces cerevisiae kinesin-related gene products required for mitotic spindle assembly. *J Cell Biol.* 118:109–120.

Hoyt, M.A., L. He, L. Totis, and W.S. Saunders. 1993. Loss of function of Saccharomyces cerevisiae kinesin-related CIN8 and KIP1 is suppressed by KAR3 motor domain mutations. *Genetics.* 135:35–44.

Huszar, D., M.E. Theoclitou, J. Skolnik, and R. Herbst. 2009. Kinesin motor proteins as targets for cancer therapy. *Cancer Metastasis Rev.* 28:197–208.

Kapitein, L.C., B.H. Kwok, J.S. Weinger, C.F. Schmidt, T.M. Kapoor, and E.J. Peterman. 2008. Microtubule cross-linking triggers the directional motility of kinesin-5. *J Cell Biol.* 182:421–428.

Kapitein, L.C., E.J. Peterman, B.H. Kwok, J.H. Kim, T.M. Kapoor, and C.F. Schmidt. 2005. The bipolar mitotic kinesin Eg5 moves on both microtubules that it crosslinks. *Nature.* 435:114–118.

Kapoor, T.M., T.U. Mayer, M.L. Coughlin, and T.J. Mitchison. 2000. Probing spindle assembly mechanisms with monastrol, a small molecule inhibitor of the mitotic kinesin, Eg5. *J Cell Biol.* 150:975–988.

Kapoor, T.M., and T.J. Mitchison. 2001. Eg5 is static in bipolar spindles relative to tubulin: evidence for a static spindle matrix. *J Cell Biol.* 154:1125–1133.

Kaseda, K., I. Crevel, K. Hirose, and R.A. Cross. 2008. Single-headed mode of kinesin-5. *EMBO Rep.* 9:761–765.

Kashina, A.S., R.J. Baskin, D.G. Cole, K.P. Wedaman, W.M. Saxton, and J.M. Scholey. 1996. A bipolar kinesin. *Nature.* 379:270–272.

Kim, E.D., R. Buckley, S. Learman, J. Richard, C. Parke, D.K. Worthylake, E.J. Wojcik, R.A. Walker, and S. Kim. 2010. Allosteric drug discrimination is coupled to mechanochemical changes in the kinesin-5 motor core. *J Biol Chem.* 285:18650–18661.

Krzysiak, T.C., and S.P. Gilbert. 2006. Dimeric Eg5 maintains processivity through alternating-site catalysis with rate-limiting ATP hydrolysis. *J Biol Chem.* 281:39444–39454.

Kwok, B.H., L.C. Kapitein, J.H. Kim, E.J. Peterman, C.F. Schmidt, and T.M. Kapoor. 2006. Allosteric inhibition of kinesin-5 modulates its processive directional motility. *Nat Chem Biol.* 2:480–485.

Lakamper, S., C. Thiede, A. Duselder, S. Reiter, M.J. Korneev, L.C. Kapitein, E.J. Peterman, and C.F. Schmidt. 2010. The effect of monastrol on the processive motility of a dimeric kinesin-5 head/kinesin-1 stalk chimera. *J Mol Biol.* 399:1–8.

Larson, A.G., N. Naber, R. Cooke, E. Pate, and S.E. Rice. 2010. The conserved L5 loop establishes the pre-powerstroke conformation of the Kinesin-5 motor, eg5. *Biophys J.* 98:2619–2627.

Lee, C.W., K. Belanger, S.C. Rao, T.M. Petrella, R.G. Tozer, L. Wood, K.J. Savage, E.A. Eisenhauer, T.W. Synold, N. Wainman, and L. Seymour. 2008. A phase II study of ispinesib (SB-715992) in patients with metastatic or recurrent malignant melanoma: a National Cancer Institute of Canada Clinical Trials Group trial. *Invest New Drugs.* 26:249–255.

Lee, Y.R., H.M. Giang, and B. Liu. 2001. A novel plant kinesin-related protein specifically associates with the phragmoplast organelles. *Plant Cell.* 13:2427–2439.

Leizerman, I., R. Avunie-Masala, M. Elkabets, A. Fich, and L. Gheber. 2004. Differential effects of monastrol in two human cell lines. *Cell Mol Life Sci.* 61:2060–2070.

Lin, S., M. Liu, O.I. Mozgova, W. Yu, and P.W. Baas. 2012. Mitotic motors coregulate microtubule patterns in axons and dendrites. *J Neurosci.* 32:14033–14049.

Liu, M., J. Ran, and J. Zhou. 2018. Non-canonical functions of the mitotic kinesin Eg5. *Thoracic Cancer.* 9:904–910.

Liu, M., X. Wang, Y. Yang, D. Li, H. Ren, Q. Zhu, Q. Chen, S. Han, J. Hao, and J. Zhou. 2010. Ectopic expression of the microtubule-dependent motor protein Eg5 promotes pancreatic tumourigenesis. *J Pathol.* 221:221–228.

Liu, X., H. Gong, and K. Huang. 2013. Oncogenic role of kinesin proteins and targeting kinesin therapy. *Cancer Sci.* 104:651–656.

Ma, N., J. Titus, A. Gable, J.L. Ross, and P. Wadsworth. 2011. TPX2 regulates the localization and activity of Eg5 in the mammalian mitotic spindle. *J Cell Biol.* 195:87–98.

Maliga, Z., and T.J. Mitchison. 2006. Small-molecule and mutational analysis of allosteric Eg5 inhibition by monastrol. *BMC Chem Biol.* 6:2.

Maliga, Z., J. Xing, H. Cheung, L.J. Juszczak, J.M. Friedman, and S.S. Rosenfeld. 2006. A pathway of structural changes produced by monastrol binding to Eg5. *J Biol Chem.* 281:7977–7982.

Mann, B.J., and P. Wadsworth. 2019. Kinesin-5 regulation and function in mitosis. *Trends Cell Biol.* 29:66–79.

Marshall, C.G., M. Torrent, O. Williams, K.A. Hamilton, and C.A. Buser. 2009. Characterization of inhibitor binding to human kinesin spindle protein by site-directed mutagenesis. *Arch. Biochem Biophys.* 484:1–7.

Mayer, T.U., T.M. Kapoor, S.J. Haggarty, R.W. King, S.L. Schreiber, and T.J. Mitchison. 1999. Small molecule inhibitor of mitotic spindle bipolarity identified in a phenotype-based screen. *Science.* 286:971–974.

Miki, T., H. Naito, M. Nishina, and G. Goshima. 2014. Endogenous localizome identifies 43 mitotic kinesins in a plant cell. *Proc Natl Acad Sci U S A.* 111:E1053–E1061.

Milic, B., A. Chakraborty, K. Han, M.C. Bassik, and S.M. Block. 2018. KIF15 nanomechanics and kinesin inhibitors, with implications for cancer chemotherapeutics. *Proc Natl Acad Sci U S A.* 115:E4613–E4622.

Miyamoto, D.T., Z.E. Perlman, K.S. Burbank, A.C. Groen, and T.J. Mitchison. 2004. The kinesin Eg5 drives poleward microtubule flux in Xenopus laevis egg extract spindles. *J Cell Biol.* 167:813–818.

Movshovich, N., V. Fridman, A. Gerson-Gurwitz, I. Shumacher, I. Gertsberg, A. Fich, M.A. Hoyt, B. Katz, and L. Gheber. 2008. Slk19-dependent mid-anaphase pause in kinesin-5-mutated cells. *J Cell Sci.* 121:2529–2539.

Muretta, J.M., W.M. Behnke-Parks, J. Major, K.J. Petersen, A. Goulet, C.A. Moores, D.D. Thomas, and S.S. Rosenfeld. 2013. Loop L5 assumes three distinct orientations during the ATPase cycle of the mitotic kinesin Eg5: a transient and time-resolved fluorescence study. *J Biol Chem.* 288:34839–34849.

Muretta, J.M., Y. Jun, S.P. Gross, J. Major, D.D. Thomas, and S.S. Rosenfeld. 2015. The structural kinetics of switch-1 and the neck linker explain the functions of kinesin-1 and Eg5. *Proc Natl Acad Sci U S A.* 112:E6606–E6613.

Muretta, J.M., B.J.N. Reddy, G. Scarabelli, A.F. Thompson, S. Jariwala, J. Major, M. Venere, J.N. Rich, B. Willard, D.D. Thomas, J. Stumpff, B.J. Grant, S.P. Gross, and S.S. Rosenfeld. 2018. A posttranslational modification of the mitotic kinesin Eg5 that enhances its mechanochemical coupling and alters its mitotic function. *Proc Natl Acad Sci U S A.* 115:E1779–E1788.

Myers, K.A., and P.W. Baas. 2007. Kinesin-5 regulates the growth of the axon by acting as a brake on its microtubule array. *J Cell Biol.* 178:1081–1091.

Nadar, V.C., A. Ketschek, K.A. Myers, G. Gallo, and P.W. Baas. 2008. Kinesin-5 is essential for growth-cone turning. *Curr Biol.* 18:1972–1977.

Pei, Y.Y., G.C. Li, J. Ran, and F.X. Wei. 2017. Kinesin family member 11 contributes to the progression and prognosis of human breast cancer. *Oncol Lett.* 14:6618–6626.

Pettersen, E.F., T.D. Goddard, C.C. Huang, G.S. Couch, D.M. Greenblatt, E.C. Meng, and T.E. Ferrin. 2004. UCSF Chimera—a visualization system for exploratory research and analysis. *J Comput Chem.* 25:1605–1612.

Rapley, J., M. Nicolas, A. Groen, L. Regue, M.T. Bertran, C. Caelles, J. Avruch, and J. Roig. 2008. The NIMA-family kinase Nek6 phosphorylates the kinesin Eg5 at a novel site necessary for mitotic spindle formation. *J Cell Sci.* 121:3912–3921.

Rath, O., and F. Kozielski. 2012. Kinesins and cancer. *Nat Rev Cancer.* 12:527–539.

Rincon, S.A., A. Lamson, R. Blackwell, V. Syrovatkina, V. Fraisier, A. Paoletti, M.D. Betterton, and P.T. Tran. 2017. Kinesin-5-independent mitotic spindle assembly requires the antiparallel microtubule crosslinker Ase1 in fission yeast. *Nat Commun.* 8:15286.

Rogers, G.C., S.L. Rogers, and D.J. Sharp. 2005. Spindle microtubules in flux. *J Cell Sci.* 118:1105–1116.

Roostalu, J., C. Hentrich, P. Bieling, I.A. Telley, E. Schiebel, and T. Surrey. 2011. Directional switching of the Kinesin cin8 through motor coupling. *Science.* 332:94–99.

Rosenfeld, S.S., J. Xing, G.M. Jefferson, and P.H. King. 2005. Docking and rolling, a model of how the mitotic motor Eg5 works. *J Biol Chem.* 280:35684–35695.

Rozelle, D.K., S.D. Hansen, and K.B. Kaplan. 2011. Chromosome passenger complexes control anaphase duration and spindle elongation via a kinesin-5 brake. *J Cell Biol.* 193:285–294.

Rusan, N.M., U.S. Tulu, C. Fagerstrom, and P. Wadsworth. 2002. Reorganization of the microtubule array in prophase/prometaphase requires cytoplasmic dynein-dependent microtubule transport. *J Cell Biol.* 158:997–1003.

Sadakane, K., M. Takaichi, and S. Maruta. 2018. Photo-control of the mitotic kinesin Eg5 using a novel photochromic inhibitor composed of a spiropyran derivative. *J Biochem.* 164:239–246.

Sarli, V., and A. Giannis. 2008. Targeting the kinesin spindle protein: basic principles and clinical implications. *Clin Cancer Res.* 14:7583–7587.

Saunders, A.M., J. Powers, S. Strome, and W.M. Saxton. 2007. Kinesin-5 acts as a brake in anaphase spindle elongation. *Curr Biol.* 17:R453–R454.

Saunders, W., V. Lengyel, and M.A. Hoyt. 1997. Mitotic spindle function in Saccharomyces cerevisiae requires a balance between different types of kinesin-related motors. *Mol Biol Cell.* 8:1025–1033.

Saunders, W.S., D. Koshland, D. Eshel, I.R. Gibbons, and M.A. Hoyt. 1995. Saccharomyces cerevisiae kinesin- and dynein-related proteins required for anaphase chromosome segregation. *J Cell Biol.* 128:617–624.

Sawin, K.E., K. LeGuellec, M. Philippe, and T.J. Mitchison. 1992. Mitotic spindle organization by a plus-end-directed microtubule motor. *Nature.* 359:540–543.

Sawin, K.E., and T.J. Mitchison. 1995. Mutations in the kinesin-like protein Eg5 disrupting localization to the mitotic spindle. *Proc Natl Acad Sci U S A.* 92:4289–4293.

Scholey, J.E., S. Nithianantham, J.M. Scholey, and J. Al-Bassam. 2014. Structural basis for the assembly of the mitotic motor Kinesin-5 into bipolar tetramers. *Elife.* 3:e02217.

Shah, J.J., J.L. Kaufman, J.A. Zonder, A.D. Cohen, W.I. Bensinger, B.W. Hilder, S.A. Rush, D.H. Walker, B.J. Tunquist, K.S. Litwiler, M. Ptaszynski, R.Z. Orlowski, and S. Lonial. 2017. A Phase 1 and 2 study of Filanesib alone and in combination with low-dose dexamethasone in relapsed/refractory multiple myeloma. *Cancer.* 123:4617–4630.

Shapira, O., and L. Gheber. 2016. Motile properties of the bi-directional kinesin-5 Cin8 are affected by phosphorylation in its motor domain. *Sci Rep.* 6:25597.

Shapira, O., A. Goldstein, J. Al-Bassam, and L. Gheber. 2017. A potential physiological role for bi-directional motility and motor clustering of mitotic kinesin-5 Cin8 in yeast mitosis. *J Cell Sci.* 130:725–734.

Sharp, D.J., K.L. McDonald, H.M. Brown, H.J. Matthies, C. Walczak, R.D. Vale, T.J. Mitchison, and J.M. Scholey. 1999a. The bipolar kinesin, KLP61F, cross-links microtubules within interpolar microtubule bundles of Drosophila embryonic mitotic spindles. *J Cell Biol.* 144:125–138.

Sharp, D.J., K.R. Yu, J.C. Sisson, W. Sullivan, and J.M. Scholey. 1999b. Antagonistic microtubule-sliding motors position mitotic centrosomes in Drosophila early embryos. *Nat Cell Biol.* 1:51–54.

Shastry, S., and W.O. Hancock. 2011. Interhead tension determines processivity across diverse N-terminal kinesins. *Proc Natl Acad Sci U S A.* 108:16253–16258.

Shimamoto, Y., S. Forth, and T.M. Kapoor. 2015. Measuring pushing and braking forces generated by ensembles of kinesin-5 crosslinking two microtubules. *Dev Cell.* 34:669–681.

Singh, S.K., H. Pandey, J. Al-Bassam, and L. Gheber. 2018. Bidirectional motility of kinesin-5 motor proteins: structural determinants, cumulative functions and physiological roles. *Cell Mol Life Sci,* 75:1757–1771.

Stoica, C., J. Park, J.M. Pare, S. Willows, and T.C. Hobman. 2010. The Kinesin motor protein Cut7 regulates biogenesis and function of Ago1-complexes. *Traffic.* 11:25–36.

Straight, A.F., J.W. Sedat, and A.W. Murray. 1998. Time-lapse microscopy reveals unique roles for kinesins during anaphase in budding yeast. *J Cell Biol.* 143:687–694.

Sturgill, E.G., S.R. Norris, Y. Guo, and R. Ohi. 2016. Kinesin-5 inhibitor resistance is driven by kinesin-12. *J Cell Biol.* 213:213–227.

Sun, L., J. Lu, Z. Niu, K. Ding, D. Bi, S. Liu, J. Li, F. Wu, H. Zhang, Z. Zhao, and S. Ding. 2015. A potent chemotherapeutic strategy with Eg5 inhibitor against gemcitabine resistant bladder cancer. *PLoS One.* 10:e0144484.

Suzuki, A., A. Gupta, S.K. Long, R. Evans, B.L. Badger, E.D. Salmon, S. Biggins, and K. Bloom. 2018. A Kinesin-5, Cin8, recruits protein phosphatase 1 to kinetochores and regulates chromosome segregation. *Curr Biol.* 28:2697–2704.e2693.

Tao, L., A. Mogilner, G. Civelekoglu-Scholey, R. Wollman, J. Evans, H. Stahlberg, and J.M. Scholey. 2006. A homotetrameric kinesin-5, KLP61F, bundles microtubules and antagonizes Ncd in motility assays. *Curr Biol.* 16:2293–2302.

Tikhonenko, I., D.K. Nag, N. Martin, and M.P. Koonce. 2008. Kinesin-5 is not essential for mitotic spindle elongation in Dictyostelium. *Cell Motil Cytoskeleton.* 65:853–862.

Tubman, E., Y. He, T.S. Hays, and D.J. Odde. 2018. Kinesin-5 mediated chromosome congression in insect spindles. *Cellular Mol Bioeng.* 11:25–36.

Turner, J., R. Anderson, J. Guo, C. Beraud, R. Fletterick, and R. Sakowicz. 2001. Crystal structure of the mitotic spindle kinesin Eg5 reveals a novel conformation of the neck-linker. *J Biol Chem.* 276:25496–25502.

Tytell, J.D., and P.K. Sorger. 2006. Analysis of kinesin motor function at budding yeast kinetochores. *JCB.* 172:861–874.

Uteng, M., C. Hentrich, K. Miura, P. Bieling, and T. Surrey. 2008. Poleward transport of Eg5 by dynein-dynactin in Xenopus laevis egg extract spindles. *J Cell Biol.* 182:715–726.

Valentine, M.T., P.M. Fordyce, T.C. Krzysiak, S.P. Gilbert, and S.M. Block. 2006. Individual dimers of the mitotic kinesin motor Eg5 step processively and support substantial loads in vitro. *Nat Cell Biol.* 8:470–476.

van den Wildenberg, S.M., L. Tao, L.C. Kapitein, C.F. Schmidt, J.M. Scholey, and E.J. Peterman. 2008. The homotetrameric kinesin-5 KLP61F preferentially crosslinks microtubules into antiparallel orientations. *Curr Biol.* 18:1860–1864.

Waitzman, J.S., A.G. Larson, J.C. Cochran, R. Naber, R. Cooke, F. Jon Kull, E. Pate, and S.E. Rice. 2011. The loop 5 element structurally and kinetically coordinates dimers of the human kinesin-5, Eg5. *Biophys J.* 101:2760–2769.

Waitzman, J.S., and S.E. Rice. 2014. Mechanism and regulation of kinesin-5, an essential motor for the mitotic spindle. *Biol Cell.* 106:1–12.

Wang, Y., X. Wu, M. Du, X. Chen, X. Ning, H. Chen, S. Wang, J. Liu, Z. Liu, R. Li, G. Fu, C. Wang, M.A. McNutt, D. Zhou, and Y. Yin. 2017. Eg5 inhibitor YL001 induces mitotic arrest and inhibits tumor proliferation. *Oncotarget.* 8:42510–42524.

Wargacki, M.M., J.C. Tay, E.G. Muller, C.L. Asbury, and T.N. Davis. 2010. Kip3, the yeast kinesin-8, is required for clustering of kinetochores at metaphase. *Cell Cycle (Georgetown, Tex.).* 9:2581–2588.

Weinger, J.S., M. Qiu, G. Yang, and T.M. Kapoor. 2011. A nonmotor microtubule binding site in kinesin-5 is required for filament crosslinking and sliding. *Curr Biol.* 21:154–160.

Whitehead, C.M., and J.B. Rattner. 1998. Expanding the role of HsEg5 within the mitotic and post-mitotic phases of the cell cycle. *J Cell Sci.* 111:2551–2561.

Wittmann, T., M. Wilm, E. Karsenti, and I. Vernos. 2000. TPX2, A novel xenopus MAP involved in spindle pole organization. *J Cell Biol.* 149:1405–1418.

Wojcik, E.J., R.S. Buckley, J. Richard, L. Liu, T.M. Huckaba, and S. Kim. 2013. Kinesin-5: cross-bridging mechanism to targeted clinical therapy. *Gene.* 531:133–149.

Yajima, J., K. Mizutani, and T. Nishizaka. 2008. A torque component present in mitotic kinesin Eg5 revealed by three-dimensional tracking. *Nat Struct Mol Biol.* 15:1119–1121.

Yan, Y., V. Sardana, B. Xu, C. Homnick, W. Halczenko, C.A. Buser, M. Schaber, G.D. Hartman, H.E. Huber, and L.C. Kuo. 2004. Inhibition of a mitotic motor protein: where, how, and conformational consequences. *J Mol Biol.* 335:547–554.

Zhang, T., S. Nirantar, H.H. Lim, I. Sinha, and U. Surana. 2009. DNA damage checkpoint maintains CDH1 in an active state to inhibit anaphase progression. *Dev Cell.* 17:541–551.

7 The Kinesin-6 Family

Claire T. Friel

CONTENTS

The Kinesin-6 family are N-terminal motor domain kinesins. Available data suggests that they are plus-end-directed translocating motors. They play a critical role in cytokinesis.

7.1 EXAMPLE FAMILY MEMBERS

Mammalian: KIF23 (MKLP1), KIF20A (MKLP2/RabKinesin-6), KIF20B (MPP1)
Drosophila melanogaster: Subito, Pavarotti
Caenorhabditis elegans: ZEN-4
Schizosaccharomyces pombe: Klp9

7.2 STRUCTURAL INFORMATION

In the Kinesin-6 family, the motor domain is located at the N-terminal end of the primary sequence (Lai et al. 2000). The motor domain is followed by a region of predicted coiled-coil (stalk region), followed by a C-terminal tail domain. The coiled-coil region mediates the interaction with various binding partners and is likely responsible for multimerisation. The yeast Kinesin-6, Klp9, has been shown to form homotetramers (Yukawa et al. 2019). As with other kinesins, the motor domain constitutes the principal site of microtubule binding, and nucleotide-dependent interaction with microtubules occurs via this domain (Atherton et al. 2017). However, in the Kinesin-6, KIF20A, a further microtubule-interaction site is found in the C-terminal half of the sequence (Echard et al. 1998). The structure of the KIF20A motor domain is consistent with that determined for other kinesin families. As for other kinesins, the major microtubule-binding site is centred on the tubulin intradimer interface, with the α4-helix located in the intradimer groove. Some loop regions within the motor domain contain Kinesin-6-specific insertions. Loop 8 contains a five-amino acid insert which is seen to contact the neighbouring microtubule protofilament in a cryogenic electron microscopy structure of KIF20A; this is the only example of this type of cross-protofilament interaction observed to date for a kinesin motor domain. Loop 6 contains a 99-amino acid insert that emerges from the side of the motor domain toward the microtubule plus-end. The position of this extended loop alters in a nucleotide-dependent manner but does not appear to contact the microtubule surface (Atherton et al. 2017).

7.3   FUNCTIONAL PROPERTIES

Members of the Kinesin-6 family studied to date are microtubule plus-end-directed translocating motors (Abaza et al. 2003, Yukawa et al. 2019), behaving in a fashion similar to a Kinesin-1. However, the Kinesin-6 KIF20A is found to have a greater affinity for microtubules in the ADP-bound state than is typical for a translocating kinesin, resulting in the difference in affinity between the so-called 'strongly bound' no-nucleotide and ATP states and the 'weakly bound' ADP state being significantly reduced (Atherton et al. 2017). KIF20A therefore retains a relatively strong interaction with the microtubule throughout its ATP turnover cycle.

Members of the Kinesin-6 family interact with various binding partners via binding regions, mainly located within the predicted coiled-coil. Both the *C. elegans* Kinesin-6, ZEN-4, and the mammalian equivalent, KIF23, form a complex with a Rho family GTPase-activating protein (GAP) via a binding region within the coiled-coil domain (Mishima, Kaitna, and Glotzer 2002). This complex, termed 'central-spindlin', has been shown to cause bundling of microtubules *in vitro*. The coiled-coil domain of Klp9 mediates interaction with the yeast microtubule crosslinker, Ase1 (Yukawa et al. 2019). Similarly, the mammalian Kinesin-6, KIF23, interacts with PRC1, the mammalian equivalent of yeast Ase1 (Kurasawa et al. 2004).

7.4   PHYSIOLOGICAL ROLE

A common theme in the physiological role of the Kinesin-6 family is their involvement in cell division. Members of this family play a crucial role in regulation of cytokinesis (Abaza et al. 2003). Knockdown of the mammalian Kinesin-6 family members, KIF23 or KIF20A, perturbs cytokinetic progression in HeLa cells, resulting in an increased proportion of multinucleated cells (Zhu et al. 2005). Knockdown of KIF20B also results in an increase in multinucleated cells, due to disruption of the timing of abscission (Janisch et al. 2018). In fission yeast, Klp9 is shown to regulate the speed of spindle elongation according to cell size, in order to maintain constant timing of mitosis (Kruger et al. 2019). In *Drosophila*, knockdown of the Kinesin-6, Pavarotti, results in defects in both mitotic spindle elongation and cleavage furrow ingression (Sommi et al. 2010). A second *Drosophila* Kinesin-6, Subito, plays a role in organisation of the meiotic spindle (Jang, Rahman, and McKim 2005).

The activity of several Kinesin-6 family members is regulated by phosphorylation. KIF20A requires phosphorylation by PLK1 to allow completion of cytokinesis (Neef et al. 2003). The interaction of Klp9 and its interacting partner Ase1 is regulated via phosphorylation, with dephosphorylation of both proteins promoting their interaction and resulting in enhanced velocity of spindle elongation (Fu et al. 2009).

7.5 INVOLVEMENT IN DISEASE

In keeping with its critical role in mitosis, overexpression of KIF20A stimulates proliferation and invasion in glioma cells and is associated with poor prognosis for glioma patients (Duan, Huang, and Shi 2016). Use of a peptide of KIF20A as a vaccine has proved effective in treatment of pancreatic cancer and has reached Phase II clinical trials (Miyazawa et al. 2017, Asahara et al. 2013).

The Kinesin-6, KIF20B, plays a crucial role in neural development. Mutation of KIF20B leads to defects in cortical neuron polarisation (McNeely et al. 2017), and a loss-of-function mutant of KIF20B in mice results in a small cerebral cortex and reduced production of neural progenitor cells (Janisch et al. 2013).

REFERENCES

Abaza, A., J. M. Soleilhac, J. Westendorf, M. Piel, I. Crevel, A. Roux, and F. Pirollet. 2003. "M phase phosphoprotein 1 is a human plus-end-directed kinesin-related protein required for cytokinesis." *J Biol Chem* 278 (30):27844–52. doi:10.1074/jbc.M304522200.

Asahara, S., K. Takeda, K. Yamao, H. Maguchi, and H. Yamaue. 2013. "Phase I/II clinical trial using HLA-A24-restricted peptide vaccine derived from KIF20A for patients with advanced pancreatic cancer." *J Transl Med* 11:291. doi:10.1186/1479-5876-11-291.

Atherton, J., I. M. Yu, A. Cook, J. M. Muretta, A. Joseph, J. Major, Y. Sourigues, J. Clause, M. Topf, S. S. Rosenfeld, A. Houdusse, and C. A. Moores. 2017. "The divergent mitotic kinesin MKLP2 exhibits atypical structure and mechanochemistry." *Elife* 6. doi:10.7554/eLife.27793.

Duan, J., W. Huang, and H. Shi. 2016. "Positive expression of KIF20A indicates poor prognosis of glioma patients." *Onco Targets Ther* 9:6741–9. doi:10.2147/OTT.S115974.

Echard, A., F. Jollivet, O. Martinez, J. J. Lacapere, A. Rousselet, I. Janoueix-Lerosey, and B. Goud. 1998. "Interaction of a Golgi-associated kinesin-like protein with Rab6." *Science* 279 (5350):580–5. doi:10.1126/science.279.5350.580.

Fu, C., J. J. Ward, I. Loiodice, G. Velve-Casquillas, F. J. Nedelec, and P. T. Tran. 2009. "Phospho-regulated interaction between kinesin-6 Klp9p and microtubule bundler Ase1p promotes spindle elongation." *Dev Cell* 17 (2):257–67. doi:10.1016/j.devcel.2009.06.012.

Jang, J. K., T. Rahman, and K. S. McKim. 2005. "The kinesinlike protein Subito contributes to central spindle assembly and organization of the meiotic spindle in Drosophila oocytes." *Mol Biol Cell* 16 (10):4684–94. doi:10.1091/mbc.e04-11-0964.

Janisch, K. M., K. C. McNeely, J. M. Dardick, S. H. Lim, and N. D. Dwyer. 2018. "Kinesin-6 KIF20B is required for efficient cytokinetic furrowing and timely abscission in human cells." *Mol Biol Cell* 29 (2):166–79. doi:10.1091/mbc.E17-08-0495.

Janisch, K. M., V. M. Vock, M. S. Fleming, A. Shrestha, C. M. Grimsley-Myers, B. A. Rasoul, S. A. Neale, T. D. Cupp, J. M. Kinchen, K. F. Liem, Jr., and N. D. Dwyer. 2013. "The vertebrate-specific Kinesin-6, Kif20b, is required for normal cytokinesis of polarized cortical stem cells and cerebral cortex size." *Development* 140 (23):4672–82. doi:10.1242/dev.093286.

Kruger, L. K., J. L. Sanchez, A. Paoletti, and P. T. Tran. 2019. "Kinesin-6 regulates cell-size-dependent spindle elongation velocity to keep mitosis duration constant in fission yeast." *Elife* 8. doi:10.7554/eLife.42182.

Kurasawa, Y., W. C. Earnshaw, Y. Mochizuki, N. Dohmae, and K. Todokoro. 2004. "Essential roles of KIF4 and its binding partner PRC1 in organized central spindle midzone formation." *EMBO J* 23 (16):3237–48. doi:10.1038/sj.emboj.7600347.

Lai, F., A. A. Fernald, N. Zhao, and M. M. Le Beau. 2000. "cDNA cloning, expression pattern, genomic structure and chromosomal location of RAB6KIFL, a human kinesin-like gene." *Gene* 248 (1–2):117–25. doi:10.1016/s0378-1119(00)00135-9.

McNeely, K. C., T. D. Cupp, J. N. Little, K. M. Janisch, A. Shrestha, and N. D. Dwyer. 2017. "Mutation of Kinesin-6 Kif20b causes defects in cortical neuron polarization and morphogenesis." *Neural Dev* 12 (1):5. doi:10.1186/s13064-017-0082-5.

Mishima, M., S. Kaitna, and M. Glotzer. 2002. "Central spindle assembly and cytokinesis require a kinesin-like protein/RhoGAP complex with microtubule bundling activity." *Dev Cell* 2 (1):41–54. doi:10.1016/s1534-5807(01)00110-1.

Miyazawa, M., M. Katsuda, H. Maguchi, A. Katanuma, H. Ishii, M. Ozaka, K. Yamao, H. Imaoka, M. Kawai, S. Hirono, K. I. Okada, and H. Yamaue. 2017. "Phase II clinical trial using novel peptide cocktail vaccine as a postoperative adjuvant treatment for surgically resected pancreatic cancer patients." *Int J Cancer* 140 (4):973–82. doi:10.1002/ijc.30510.

Neef, R., C. Preisinger, J. Sutcliffe, R. Kopajtich, E. A. Nigg, T. U. Mayer, and F. A. Barr. 2003. "Phosphorylation of mitotic kinesin-like protein 2 by polo-like kinase 1 is required for cytokinesis." *J Cell Biol* 162 (5):863–75. doi:10.1083/jcb.200306009.

Sommi, P., R. Ananthakrishnan, D. K. Cheerambathur, M. Kwon, S. Morales-Mulia, I. Brust-Mascher, and A. Mogilner. 2010. "A mitotic kinesin-6, Pav-KLP, mediates interdependent cortical reorganization and spindle dynamics in Drosophila embryos." *J Cell Sci* 123 (Pt 11):1862–72. doi:10.1242/jcs.064048.

Yukawa, M., M. Okazaki, Y. Teratani, K. Furuta, and T. Toda. 2019. "Kinesin-6 Klp9 plays motor-dependent and -independent roles in collaboration with Kinesin-5 Cut7 and the microtubule crosslinker Ase1 in fission yeast." *Sci Rep* 9 (1):7336. doi:10.1038/s41598-019-43774-7.

Zhu, C., J. Zhao, M. Bibikova, J. D. Leverson, E. Bossy-Wetzel, J. B. Fan, R. T. Abraham, and W. Jiang. 2005. "Functional analysis of human microtubule-based motor proteins, the kinesins and dyneins, in mitosis/cytokinesis using RNA interference." *Mol Biol Cell* 16 (7):3187–99. doi:10.1091/mbc.e05-02-0167.

8 The Kinesin-8 Family

Tianyang Liu, Alejandro Peña,
Fiona Shilliday, and Carolyn A. Moores

CONTENTS

The Kinesin-8 family are dual-function motors that both step processively toward microtubule plus ends and depolymerise/destabilise microtubules.

8.1   EXAMPLE FAMILY MEMBERS

Vertebrates: KIF18A, KIF18B, KIF19
Drosophila melanogaster: KLP67A
Caenorhabditis elegans: KLP-13
Saccharomyces cerevisiae: Kip3
Schizosaccharomyces pombe: KLP5/6

8.2   STRUCTURAL INFORMATION

Kinesin-8 protein sequences can be divided into three distinct regions: an N-terminal motor domain (MD), connected by a neck linker (NL) to a stalk domain involved in dimerisation, and a C-terminal tail domain (Figure 8.1A) (Goldstein 2001, Verhey and Hammond 2009). Binding sites for accessory proteins (importin-α, mitotic centromere-associated kinesin (MCAK), EB1) that control Kinesin-8 activity and localisation have also been identified (Stout et al. 2011, Tanenbaum et al. 2011). Structures of the motor region of Kif18A and Kif19 have been determined using X-ray crystallography, while lower-resolution structures of KIF18A/KIF19 motor-microtubule (MT) complexes have also been calculated using cryo-electron microscopy (cryo-EM) and image reconstruction (Table 8.1, online) (Peters et al. 2010, Locke et al. 2017, Wang et al. 2016). No structures of the stalk and/or C-terminal domains from any Kinesin-8s are currently available.

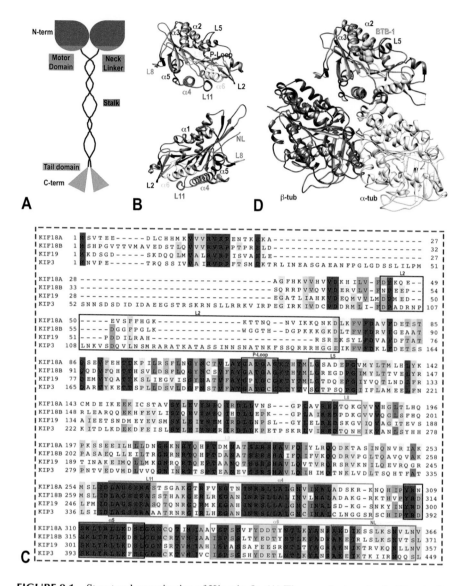

FIGURE 8.1 Structural organisation of Kinesin-8s. (A) Three main domains in Kinesin-8s: the motor domain (dark grey) followed by the neck linker peptide (light grey), which connects to the coiled-coil dimerisation domain (stalk) and the C-terminal tail domain. (B) Cryo-EM-derived model of KIF18A_MDNL in the presence of adenylyl-imidodiphosphate (AMP-PNP) with key regions labelled. (C) Alignment of motor domain/neck linker sequences from human KIF18A, KIF18B and KIF19 and budding yeast Kip3, using T-coffee (Notredame, Higgins, and Heringa 2000), with functionally significant regions highlighted. (D) KIF18A_MDNL-microtubule-bound complex in the presence of BTB-1 (shown in space filling).

As with all kinesins, the 40 kD Kinesin-8 MD is composed of a central β-sheet sandwiched between two sets of three α-helices (Figure 8.1B). Helices-α4, -α5 and -α6 lie at the motor-MT interface, while helices-α1, -α2 and -α3 face away from the MT. Helices-α4 and -α5, together with loop-2 (which is particularly extended in Kinesin-8s), loop-8, loop-11 and loop-12, form the MT-binding surface. Several of these MT-binding elements are also conserved across the superfamily, while others are more family-specific. Conservation among kinesins is particularly high in the nucleotide-binding motifs: the P-loop, switch-I (loop-9) and switch-II (loop-11) (Figure 8.1C). The ~20-amino acid NL emerges from the end of helix-α6 and mediates communication within the dimeric motor to enable processive stepping of Kinesin-8s along the MT lattice (see Section 8.3 Functional Properties below).

The X-ray structure of human KIF18A_MD bound to Mg_2-ADP was the first high-resolution structure of a Kinesin-8 to be determined, and the orientation of helix-α4 with respect to the rest of the MD is similar compared to other ADP-bound kinesins (Peters et al. 2010, Sablin et al. 1996, Kull et al. 1996). As is sometimes seen in X-ray-derived MD structures, loops that are involved in MT-binding are partially disordered (Peters et al. 2010, Woehlke et al. 1997, Shipley et al. 2004, Ogawa et al. 2017). The earliest cryo-EM structures of MT-bound human KIF18A_MD were determined in the absence of nucleotide (no-nucleotide, NN) and in the presence of adenylyl-imidodiphosphate (AMP-PNP), a non-hydrolysable ATP analogue at ~10 Å resolution (Peters et al. 2010). In both of these nucleotide states, helix-α4 adopts an extended conformation at the αβ-tubulin intradimer interface, while the loops that were disordered in the crystal structure became structured and therefore visible, in contact with the MT. However, apart from the presence or absence of density corresponding to a bound nucleotide, the conformation of KIF18A_MD in the NN and AMP-PNP structures was essentially the same, suggesting that the NL is vital to mediate a functionally relevant response of the motor to nucleotide-binding.

This conclusion was supported by a subsequent, higher resolution study (~7 Å) of MT-bound KIF18A_MDNL, in which the motor was observed to undergo nucleotide-dependent conformational changes (Locke et al. 2017). In the MT-bound NN structure, the empty conformation of the nucleotide-binding pocket is clear, while the relative position of helix-α4/-α6 prevents NL docking, which is thus flexible and not visible. This conformation of the motor is very similar to those of the MT-bound KIF18A_MD structures (Peters et al. 2010). However, in the presence of AMP-PNP, KIF18A_MDNL undergoes a conformational change. Loops at the nucleotide-binding site close around the bound AMPPNP while allosterically, an extension of helix-α6 and the docking of NL along the MD towards the plus-end of the MT is observed (Locke et al. 2017), consistent with plus-end directed movement in the context of the Kinesin-8 dimer. Overall, the conformational changes seen in the transition between the NN and AMPPNP states of KIF18A_MDNL are consistent with the previously described movements of subdomains within other kinesin MDs, which respond in a coordinated manner to MT and nucleotide-binding to drive motility (Cao et al. 2014, Shang et al. 2014, Atherton et al. 2014, Goulet et al. 2012, Atherton et al. 2017).

A third MT-bound KIF18A_MDNL cryo-EM structure was determined in the presence of BTB-1, a KIF18A-specific small-molecule inhibitor (Catarinella et al. 2009) that traps the motor on the MT (Figure 8.1D). The overall conformation of BTB-1-bound KIF18A_MDNL is very similar to its NN state. Although the resolution of the reconstruction was not sufficient to visualise the bound small molecule directly, computational analyses identified a putative BTB-1-binding site between helices-α2 and -α3, which was also validated by point mutagenesis (Locke et al. 2017). BTB-1 binds at the junction of two subdomains within the MD, thereby locking KIF18A_MDNL in the MT-bound state and blocking subdomain rearrangements that are required for motor function.

Mouse KIF19_MDNL has been crystallised bound to Mg_2-ADP (Wang et al. 2016). Relative to the KIF18A crystal structure, the KIF19 structure shows partial disordering and a large dislocation of helix-α4-α5 and loop-12 away from the core β-sheet. In addition, loop-2 is longer and more ordered in KIF19 compared with KIF18A, while helix-α6 is shorter and the KIF19 NL is not visualised. Cryo-EM of MT-bound NN KIF19A_MDNL revealed an empty nucleotide-binding pocket, with helix-α4-α5 and loop-12 now packed against the core β-sheet and lying at the motor-MT interface. Loop-2 retains the same extended structure as seen in the crystal structure and contacts the MT surface. Overall, this conformation of MT-bound NN KIF19_MDNL is very similar to that of MT-bound NN KIF18A_MDNL.

As with other kinesins, formation of the Kinesin-8-MT complex stimulates ADP release and allows structural coupling within the MD, such that the occupancy of the nucleotide-binding site controls the conformation of the NL. ATP-binding stimulates reorientation of the NL towards the MT plus-end, supporting plus-end-directed stepping through alternating head binding/unbinding in the context of the dimer. However, the function of Kinesin-8 motors involves both walking towards MT plus-ends and regulating the dynamics of these ends on arrival. Structural studies have also begun to shed light – albeit at low resolution – on the mechanistic basis of these two distinct activities. As well as binding along the MT lattice, KIF18A_MD and KIF18A_MDNL also induce formation of tubulin rings in the presence of AMP-PNP at MT ends, which have been structurally characterised in 2D using electron microscopy (EM) (Peters et al. 2010, Locke et al. 2017). These curved protofilament-motor complexes are reminiscent of the products formed by depolymerising Kinesin-13s in the presence of AMP-PNP (Moores et al. 2002, Benoit, Asenjo, and Sosa 2018, Tan, Rice, and Sosa 2008) and show that the motor-tubulin interaction in these complexes is very similar to that seen on the lattice, albeit that the tubulin is curved. Further work is required to define the structural mechanism by which Kinesin-8s influence MT dynamics at their ends.

8.3 FUNCTIONAL PROPERTIES:

Kinesin-8s are ATP-driven dual-function motors. Overall, their functions are related to regulating MT length and dynamics rather than transporting cargo *per se*, although many Kinesin-8s can step processively towards the plus-ends of MTs. Budding yeast Kip3 and KIF18A have also been demonstrated to have MT depolymerase activity, whereas, as with fission yeast Klp5/6, these proteins also increase rescue and catastrophe frequencies and the growth rate of MTs (Mayr et al. 2007, Varga et al. 2006, Varga et al. 2009, Unsworth et al. 2008). Furthermore, both Kip3 and KIF18A can reduce growth and shrinkage duration and distance as well as increasing pause duration (Du, English, and Ohi 2010). In other words, Kinesin-8s exhibit both MT-destabilising and -stabilising activities, which enable them to constrain the extent of MT growth and shrinkage, lowering the overall levels of MT dynamicity (Stumpff et al. 2008, Du, English, and Ohi 2010, Fukuda et al. 2014).

Studies of the molecular properties of dimeric full-length Kinesin-8s typically show that these proteins are highly processive (>5 µm run length). Their motility is slow and they exhibit relatively long plus-end dwell times which depend on their tail domains (see Table 8.2 online) (Varga et al. 2009, Mayr et al. 2011, Mockel et al. 2017, Niwa et al. 2012, Wang, Nitta, et al. 2016, Arellano-Santoyo et al. 2017, Varga et al. 2006, Stumpff et al. 2011). In several Kinesin-8s, these C-terminal domains bind directly to MTs, providing an additional site of MT tethering (Stout et al. 2011, Weaver et al. 2011, Su et al. 2011, Mayr et al. 2011, Stumpff et al. 2011) and also enable MT crosslinking (Su et al. 2013). Using single-molecule imaging *in vitro*, Kip3 was observed walking to MT plus-ends – sometimes switching the protofilament along which it walks (Bugiel, Böhl, and Schäffer 2015) – and then depolymerising these MTs in a length-dependent manner (Varga et al. 2009, Varga et al. 2006). The antenna model was proposed to explain Kip3's length-dependent MT depolymerisation: because of its high processivity, Kip3 forms a concentration gradient towards MT plus-ends that is proportional to MT length; therefore, longer MTs have higher concentrations of Kip3 at their plus-ends and thus faster depolymerisation rates. This model provides an explanation for Kip3's ability to regulate the overall length of MTs in a population, but whether the antenna model is relevant to other Kinesin-8s has not yet been demonstrated. In fact, whether all Kinesin-8s are depolymerases is still in question. For example, human KIF18A was reported as a MT depolymerase in some studies (Mayr et al. 2007, Peters et al. 2010, Locke et al. 2017), but no depolymerisation activity was observed in others (Du, English, and Ohi 2010). Rather, its inhibition of MT polymerisation was proposed to lead to reduced dynamicity of MTs, an activity that is more consistent with its cellular functions (Du, English, and Ohi 2010).

Monomeric Kinesin-8 MD constructs cannot take processive steps but functional studies focusing on monomers shed light on the underlying properties of Kinesin-8 MDs and have complemented structural studies. Comparison of the steady-state ATPase rates and gliding activities show that the KIF18A_MD construct is slower than KIF18A_MDNL and has lower MT affinity. Both constructs exhibit ATP-dependent MT depolymerisation from plus and minus ends and induce the formation of tubulin rings in the presence of AMP-PNP. All these *in vitro* activities of monomeric KIF18A constructs were inhibited by the small molecule BTB-1 (Locke et al. 2017). A monomeric mouse KIF19A construct was also observed to drive ATP-dependent motility and depolymerase activity (Wang et al. 2016). Overall, many of the properties of full-length Kinesin-8s reflect the intrinsic properties of their individual MDs.

Stepping by dimeric Kinesin-8 along MTs is considered to use a canonical alternating head mechanism. However, different models have been proposed for the mechanism by which they regulate MT dynamics at their ends (Varga et al. 2009, Arellano-Santoyo et al. 2017, Su et al. 2011). The "bump-off model" for Kip3 proposed that molecules of motor dimers arriving at the MT plus-end promote the dissociation of the most distal Kip3 molecule, accompanied by removal of one or two tubulin dimers (Varga et al. 2009). This model depends on processive stepping of dimeric motors, but monomeric Kinesin-8s can also influence MT dynamics, suggesting that other mechanisms must be operating. In the "two-state binding switch model", the Kip3 motor binds tightly to the more curved tubulin subunits found at the MT ends, which thereby inhibits their ATPase activity and leads to catastrophe or MT disassembly (Arellano-Santoyo et al. 2017). This model is more consistent with the ability of both monomeric and dimeric Kinesin-8s constructs to influence MT dynamics; in this context, the plus-end specific activity of Kinesin-8 dimers is explained by these constructs first stepping to the MT plus-end, whereas monomeric proteins can bind to either end of the MT, thereby influencing MT length at both plus and minus-ends. A similar mechanism was also reported for the ciliary Kinesin-8, KIF19 (Wang et al. 2016). The particular role of the characteristic extended loop-2 in Kinesin-8s in their depolymerase activity has been investigated, both because this region interacts with the MT (Wang et al. 2016, Peters et al. 2010, Locke et al. 2017), and because loop-2 is critical in the depolymerase activity of Kinesin-13s (Ogawa et al. 2004, Shipley et al. 2004). However, while loop-2 residues are vital for KIF19 depolymerase activity (Wang et al. 2016), loop-2 in KIF18A (Kim, Fonseca, and Stumpff 2014) and Kip3 (Arellano-Santoyo et al. 2017) does not appear to be directly involved in MT depolymerisation activity, and contributes instead to motor processivity and/or MT end-binding.

The origins of differences in Kinesin-8 properties across species are still not well understood. *In vitro* work, combined with structural studies of a wider range of Kinesin-8s, will deepen our mechanistic insight concerning these motors. Furthermore, the models derived from *in vitro* data can account for Kinesin-8 motility and destabilising activity, but the Kinesin-8-dependent activities that emerge in cellular settings – especially their MT-stabilising effect – likely arise from more complex molecular behaviours. The antenna model for Kip3 can provide some rationalisation for these observations: Kip3 accumulation at growing MT plus-ends induces catastrophe – incoming Kip3s do not accumulate at the shrinking plus-end (Gupta et al. 2006), and their reduced concentrations allow MT rescue (Su et al. 2011). However, the requirement by Kinesin-8s for collaboration with other +TIPs (MT plus-end tracking proteins) to execute their complex regulation of MT dynamics is also still not understood (Tanenbaum et al. 2011, Stout et al. 2011, Sanchez-Perez et al. 2005), and would benefit from *in vitro* multi-component reconstitution analyses of these MT regulators

8.4 PHYSIOLOGICAL ROLES

Kinesin-8s are widely expressed in eukaryotes and their best-studied physiological role is the regulation of MT dynamics during cell division. The precise contribution(s) of each Kinesin-8 to the complex rearrangements that take place during mitosis and meiosis depends on the specific spindle context within which they operate. Nevertheless, common themes are emerging in the roles of these multi-tasking motors in a range of eukaryotes.

In dividing budding yeast, Kip3 is required for mitosis spindle positioning, to align the spindle along the mother–daughter axis, to cluster kinetochores and to maintain mitotic arrest in response to mis-positioned spindles (Fukuda et al. 2014, Varga et al. 2006, Gupta et al. 2006, Varga et al. 2009). Regulation by Kip3 of MT plus-end dynamics is involved in all these activities, while MT crosslinking and sliding also contributes to spindle elongation in anaphase. Fission yeast expresses two Kinesin-8s, Klp5 and Klp6, that work together to regulate MT dynamics in complex ways. They both play essential roles in meiosis and are involved in normal chromosome alignment and segregation during mitosis (West, Malmstrom, and McIntosh 2001, Unsworth et al. 2008). *D. melanogaster* Klp67A is required for maintenance of spindle bipolarity and spindle MT length, and disruption of its function prevents completion of mitosis (Goshima and Vale 2003, Savoian and Glover 2010) and meiosis (Savoian et al. 2004). It localises to kinetochore-attached MTs (kMTs) and controls their length and attachment to kinetochores (Edzuka and Goshima 2019).

KIF18A is the best-studied mammalian Kinesin-8 and has a very broad tissue expression pattern, reflecting its importance in mitosis. KIF18A is found at the plus-ends of kMTs (Figure 8.2) and contributes to chromosome congression (Zhu et al. 2005, Mayr et al. 2007, Stumpff et al. 2008). Its depletion produces mitotic spindles that are abnormally long, and which exhibit a loss of tension across sister kinetochores. In KIF18A-depleted cells, the chromosome oscillations that precede metaphase are exaggerated, demonstrating that KIF18A is important for regulating these movements. It is also involved in controlling poleward chromosome movement in anaphase (Stumpff et al. 2008). Lack of KIF18A in mammalian germ cells and some tumour cell lines activates the spindle assembly checkpoint, ultimately leading to cell death (Czechanski et al. 2015, Janssen et al. 2018). In contrast, in primary mouse embryonic fibroblasts and presumably many other cells in a mouse line with mutated KIF18A, mitosis proceeds in the absence of KIF18A, and the resulting misaligned chromosomes have no effect on daughter cell ploidy (Czechanski et al. 2015). However, as a result of greater dispersion of chromosomes during anaphase in the absence of KIF18A, aberrant chromosomal organisation within daughter nuclei

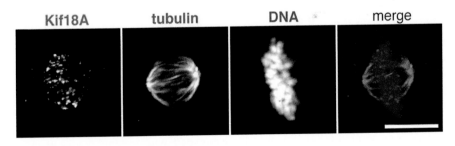

FIGURE 8.2 Localisation of KIF18A at the plus-end of kinetochore microtubules. HeLa cells in mitosis were imaged by immunofluorescence using 4',6-diamidino-2-phenylindole (DAPI)/DNA (blue) and antibodies against KIF18A (red) and α-tubulin (green). Scale bar = 10 μm. Reproduced with permission from Mayr et al. (2007).

is observed, together with micronuclei formation. Over time, these effects slow cell proliferation and have been linked to compromised growth and reduced survival of these mutant mice (Fonseca et al. 2019).

KIF18A has a C-terminal nuclear localisation sequence (NLS) (Du, English, and Ohi 2010). KIF18A activity is modulated by the antagonistic actions of cyclin-dependent kinase (Cdk1) and protein phosphatase-1 (PP1). Phosphorylation in the protein C-terminus inhibits KIF18A activity, thereby promoting chromosome oscillations, while chromosome attachment promotes recruitment of PP1, activating KIF18A and dampening chromosome movements (Häfner et al. 2014). KIF18A activity within the spindle is also regulated via sequestration by kinesin-binding protein (KBP) (Malaby et al. 2019). Studies of KBP regulatory activity also showed that KIF18A regulates MT dynamics in neurons (Kevenaar et al. 2016).

A second Kinesin-8 member, KIF18B, has a role in mitosis distinct from KIF18A, and is specifically involved in regulating astral MTs. Mammalian KIF18B localises to astral MT plus-ends where it regulates their length. Loss of KIF18B function produces long astral MTs and positional instability of the spindle in dividing cells (Tanenbaum et al. 2011, Walczak et al. 2016, McHugh, Gluszek, and Welburn 2018). At the plus-ends of astral MTs, KIF18B interacts with both EB1 (Tanenbaum et al. 2011, Stout et al. 2011) and the Kinesin-13 MCAK. MCAK localisation is required for MT depolymerisation activity at these MT ends and is negatively regulated by Aurora kinases via phosphorylation of MCAK (Tanenbaum et al. 2011, McHugh, Gluszek, and Welburn 2018).

Beyond roles in cell division, the mammalian Kinesin-8, KIF19, has roles in regulating the length of motile cilia. These MT-based organelles are found in various tissues and have a critical role in generating fluid flow around these tissues, which is vital for their development and maintenance. Length control of cilia is important for their diverse function. KIF19 localises at the tips of cilia and KIF19 knockout mice exhibit aberrantly elongated and thus dysfunctional cilia, and exhibit hydrocephalus and female infertility (Niwa et al. 2012). Phylogenetic analysis (Wickstead, Gull, and Richards 2010) indicates that Kinesin-8s are found in a wide range of other eukaryotes but so far little is known about the properties or roles of these motors.

8.5 INVOLVEMENT IN DISEASE

Perhaps due to its ubiquitous expression in most tissues, human Kinesin-8, KIF18A, is associated with a range of cancer types. Its mRNA is upregulated in colorectal cancer and is linked to increased cell proliferation and migration in cancerous tissue, and overall tumour progression and metastasis. Consistently, *in vitro* gene silencing of *Kif18A* results in reduced cell proliferation, migration and invasion (Nagahara et al. 2011). In mice, KIF18A overexpression enhances inflammation associated with colorectal tumours by upregulating PI3kinase-Akt signalling (Zhu et al. 2005).

Increased levels of KIF18A are also associated with tumour grade, metastasis and poor survival in breast cancer (Zhang et al. 2010, Kasahara et al. 2016). Upregulation of KIF18A mRNA is observed in breast cancer tissue together with a two- to five-fold increase of KIF18A protein. This abnormally high expression of KIF18A leads to creation of multinucleate cells exhibiting aneuploidy or polyploidy

(Zhang et al. 2010), but can also lead to an increase in cellular replication and tumour growth acceleration (Kasahara et al. 2016). Similar to colorectal cancer, knockdown of KIF18A inhibits cell proliferation and tumorigenesis, and its loss in tumour cells induces anoikis because of loss of cell–matrix interactions (Zhang et al. 2010). Raised levels of KIF18A are significantly associated with lymph node metastasis, and patients with higher KIF18A expression have a poorer disease prognosis (Kasahara et al. 2016). KIF18A has also been associated with hepatocellular carcinoma (Liao et al. 2014, Luo et al. 2018), renal cell carcinoma (Chen et al. 2016) and ovarian cancer (Schiewek et al. 2018). Conversely, KIF18A expression is reduced in gastric cancer tissue, indicating that it might be a protective factor in this tissue (Wang et al. 2016).

Inhibition of KIF18A has been investigated as an anti-cancer therapy, with the compound BTB-1 identified as a specific inhibitor of KIF18A *in vitro* (Catarinella et al. 2009). However, BTB-1 and related compounds were found to affect MT polymerisation in cells, an undesirable off-target effect that increased cytotoxicity (Braun et al. 2015). Therefore, while the anti-mitotic potential of specifically inhibiting KIF18A by small molecules remains, further work is required.

Human KIF18B overexpression is also associated with some cancer types. It is over-expressed in cervical cancer tissues and is associated with clinical factors such as increased tumour size and histological grade. Overexpression increases cell proliferation, migration and invasion while knockdown inhibits these activities, resulting in G1 cell cycle arrest. Expression silencing also reduces levels of a number of signalling proteins upstream of the Wnt/β-catenin pathway, which affects tumour cell metastasis (Wu et al. 2018). Over-expression also causes significant dysregulation of other cell cycle regulatory genes, leading KIF18B to be classified in some cases as a "cancer-driver" (Itzel et al. 2015).

REFERENCES

Arellano-Santoyo, H., E. A. Geyer, E. Stokasimov, G. Y. Chen, X. Su, W. Hancock, L. M. Rice, and D. Pellman. 2017. "A tubulin binding switch underlies Kip3/Kinesin-8 depolymerase activity." *Dev Cell* 42 (1):37–51 e8. doi:10.1016/j.devcel.2017.06.011.

Atherton, J., I. Farabella, I. M. Yu, S. S. Rosenfeld, A. Houdusse, M. Topf, and C. A. Moores. 2014. "Conserved mechanisms of microtubule-stimulated ADP release, ATP binding, and force generation in transport kinesins." *Elife* 3:e03680. doi:10.7554/eLife.03680.

Atherton, J., I. M. Yu, A. Cook, J. M. Muretta, A. Joseph, J. Major, Y. Sourigues, J. Clause, M. Topf, S. S. Rosenfeld, A. Houdusse, and C. A. Moores. 2017. "The divergent mitotic kinesin MKLP2 exhibits atypical structure and mechanochemistry." *Elife* 6. doi:10.7554/eLife.27793.

Benoit, M. P. M. H., A. B. Asenjo, and H. Sosa. 2018. "Cryo-EM reveals the structural basis of microtubule depolymerization by kinesin-13s." *Nat Commun* 9 (1):1662. doi:10.1038/s41467-018-04044-8.

Braun, J., M. M. Mockel, T. Strittmatter, A. Marx, U. Groth, and T. U. Mayer. 2015. "Synthesis and biological evaluation of optimized inhibitors of the mitotic kinesin Kif18A." *ACS Chem Biol* 10 (2):554–60. doi:10.1021/cb500789h.

Bugiel, M., E. Böhl, and E. Schäffer. 2015. "The Kinesin-8 Kip3 switches protofilaments in a sideward random walk asymmetrically biased by force." *Biophys J* 108 (8):2019–27. doi:10.1016/j.bpj.2015.03.022.

Cao, L., W. Wang, Q. Jiang, C. Wang, M. Knossow, and B. Gigant. 2014. "The structure of apo-kinesin bound to tubulin links the nucleotide cycle to movement." *Nat Commun* 5:5364. doi:10.1038/ncomms6364.

Catarinella, M., T. Grüner, T. Strittmatter, A. Marx, and T. U. Mayer. 2009. "BTB-1: a small molecule inhibitor of the mitotic motor protein Kif18A." *Angew Chem Int Ed Engl* 48 (48):9072–6. doi:10.1002/anie.200904510.

Chen, Q. I., B. Cao, N. Nan, Y. U. Wang, X. U. Zhai, Y. Li, and T. Chong. 2016. "Elevated expression of KIF18A enhances cell proliferation and predicts poor survival in human clear cell renal carcinoma." *Exp Ther Med* 12 (1):377–83. doi:10.3892/etm.2016.3335.

Czechanski, A., H. Kim, C. Byers, I. Greenstein, J. Stumpff, and L. G. Reinholdt. 2015. "Kif18a is specifically required for mitotic progression during germ line development." *Dev Biol* 402 (2):253–62. doi:10.1016/j.ydbio.2015.03.011.

Du, Y., C. A. English, and R. Ohi. 2010. "The kinesin-8 Kif18A dampens microtubule plus-end dynamics." *Curr Biol* 20 (4):374–80. doi:10.1016/j.cub.2009.12.049.

Edzuka, T., and G. Goshima. 2019. "Kinesin-8 stabilizes the kinetochore-microtubule interaction." *J Cell Biol* 218 (2):474–88. doi:10.1083/jcb.201807077.

Fonseca, C. L., H. L. H. Malaby, L. A. Sepaniac, W. Martin, C. Byers, A. Czechanski, D. Messinger, M. Tang, R. Ohi, L. G. Reinholdt, and J. Stumpff. 2019. "Mitotic chromosome alignment ensures mitotic fidelity by promoting interchromosomal compaction during anaphase." *J Cell Biol* 218 (4):1148–63. doi:10.1083/jcb.201807228.

Fukuda, Y., A. Luchniak, E. R. Murphy, and M. L. Gupta, Jr. 2014. "Spatial control of microtubule length and lifetime by opposing stabilizing and destabilizing functions of kinesin-8." *Curr Biol* 24 (16):1826–35. doi:10.1016/j.cub.2014.06.069.

Goldstein, L. S. 2001. "Molecular motors: from one motor many tails to one motor many tales." *Trends Cell Biol* 11 (12):477–82.

Goshima, G., and R. D. Vale. 2003. "The roles of microtubule-based motor proteins in mitosis: comprehensive RNAi analysis in the Drosophila S2 cell line." *J Cell Biol* 162 (6):1003–16. doi:10.1083/jcb.200303022.

Goulet, A., W. M. Behnke-Parks, C. V. Sindelar, J. Major, S. S. Rosenfeld, and C. A. Moores. 2012. "The structural basis of force generation by the mitotic motor kinesin-5." *J Biol Chem* 287 (53):44654–66. doi:10.1074/jbc.M112.404228.

Gupta, M. L., Jr., P. Carvalho, D. M. Roof, and D. Pellman. 2006. "Plus end-specific depolymerase activity of Kip3, a kinesin-8 protein, explains its role in positioning the yeast mitotic spindle." *Nat Cell Biol* 8 (9):913–23. doi:10.1038/ncb1457.

Häfner, J., M. I. Mayr, M. M. Möckel, and T. U. Mayer. 2014. "Pre-anaphase chromosome oscillations are regulated by the antagonistic activities of Cdk1 and PP1 on Kif18A." *Nat Commun* 5:4397. doi:10.1038/ncomms5397.

Itzel, T., P. Scholz, T. Maass, M. Krupp, J. U. Marquardt, S. Strand, D. Becker, F. Staib, H. Binder, S. Roessler, X. W. Wang, S. Thorgeirsson, M. Muller, P. R. Galle, and A. Teufel. 2015. "Translating bioinformatics in oncology: guilt-by-profiling analysis and identification of KIF18B and CDCA3 as novel driver genes in carcinogenesis." *Bioinformatics* 31 (2):216–24. doi:10.1093/bioinformatics/btu586.

Janssen, L. M. E., T. V. Averink, V. A. Blomen, T. R. Brummelkamp, R. H. Medema, and J. A. Raaijmakers. 2018. "Loss of Kif18A results in spindle assembly checkpoint activation at microtubule-attached kinetochores." *Curr Biol* 28 (17):2685–96 e4. doi:10.1016/j.cub.2018.06.026.

Kasahara, M., M. Nagahara, T. Nakagawa, T. Ishikawa, T. Sato, H. Uetake, and K. Sugihara. 2016. "Clinicopathological relevance of kinesin family member 18A expression in invasive breast cancer." *Oncol Lett* 12 (3):1909–14. doi:10.3892/ol.2016.4823.

Kevenaar, J. T., S. Bianchi, M. van Spronsen, N. Olieric, J. Lipka, C. P. Frias, M. Mikhaylova, M. Harterink, N. Keijzer, P. S. Wulf, M. Hilbert, L. C. Kapitein, E. de Graaff, A. Ahkmanova, M. O. Steinmetz, and C. C. Hoogenraad. 2016. "Kinesin-binding protein controls microtubule dynamics and cargo trafficking by regulating kinesin motor activity." *Curr Biol* 26 (7):849–61. doi:10.1016/j.cub.2016.01.048.

Kim, H., C. Fonseca, and J. Stumpff. 2014. "A unique kinesin-8 surface loop provides specificity for chromosome alignment." *Mol Biol Cell* 25 (21):3319–29. doi:10.1091/mbc.E14-06-1132.

Kull, F. J., E. P. Sablin, R. Lau, R. J. Fletterick, and R. D. Vale. 1996. "Crystal structure of the kinesin motor domain reveals a structural similarity to myosin." *Nature* 380 (6574):550–5. doi:10.1038/380550a0.

Liao, W., G. Huang, Y. Liao, J. Yang, Q. Chen, S. Xiao, J. Jin, S. He, and C. Wang. 2014. "High KIF18A expression correlates with unfavorable prognosis in primary hepatocellular carcinoma." *Oncotarget* 5 (21):10271–9. doi:10.18632/oncotarget.2082.

Locke, J., A. P. Joseph, A. Pena, M. M. Mockel, T. U. Mayer, M. Topf, and C. A. Moores. 2017. "Structural basis of human kinesin-8 function and inhibition." *Proc Natl Acad Sci U S A* 114 (45):E9539–48. doi:10.1073/pnas.1712169114.

Luo, W., M. Liao, Y. Liao, X. Chen, C. Huang, J. Fan, and W. Liao. 2018. "The role of kinesin KIF18A in the invasion and metastasis of hepatocellular carcinoma." *World J Surg Oncol* 16 (1):36. doi:10.1186/s12957-018-1342-5.

Malaby, H. L. H., M. E. Dumas, R. Ohi, and J. Stumpff. 2019. "Kinesin-binding protein ensures accurate chromosome segregation by buffering KIF18A and KIF15." *J Cell Biol* 218 (4):1218–34. doi:10.1083/jcb.201806195.

Mayr, M. I., S. Hummer, J. Bormann, T. Gruner, S. Adio, G. Woehlke, and T. U. Mayer. 2007. "The human kinesin Kif18A is a motile microtubule depolymerase essential for chromosome congression." *Curr Biol* 17 (6):488–98. doi:10.1016/j.cub.2007.02.036.

Mayr, M. I., M. Storch, J. Howard, and T. U. Mayer. 2011. "A non-motor microtubule binding site is essential for the high processivity and mitotic function of kinesin-8 Kif18A." *PLoS One* 6 (11):e27471. doi:10.1371/journal.pone.0027471.

McHugh, T., A. A. Gluszek, and J. P. I. Welburn. 2018. "Microtubule end tethering of a processive kinesin-8 motor Kif18b is required for spindle positioning." *J Cell Biol* 217 (7):2403–16. doi:10.1083/jcb.201705209.

Mockel, M. M., A. Heim, T. Tischer, and T. U. Mayer. 2017. "Xenopus laevis Kif18A is a highly processive kinesin required for meiotic spindle integrity." *Biol Open* 6 (4):463–70. doi:10.1242/bio.023952.

Moores, C. A., M. Yu, J. Guo, C. Beraud, R. Sakowicz, and R. A. Milligan. 2002. "A mechanism for microtubule depolymerization by KinI kinesins." *Mol Cell* 9 (4):903–9.

Nagahara, M., N. Nishida, M. Iwatsuki, S. Ishimaru, K. Mimori, F. Tanaka, T. Nakagawa, T. Sato, K. Sugihara, D. S. Hoon, and M. Mori. 2011. "Kinesin 18A expression: clinical relevance to colorectal cancer progression." *Int J Cancer* 129 (11):2543–52. doi:10.1002/ijc.25916.

Niwa, S., K. Nakajima, H. Miki, Y. Minato, D. Wang, and N. Hirokawa. 2012. "KIF19A is a microtubule-depolymerizing kinesin for ciliary length control." *Dev Cell* 23 (6):1167–75. doi:10.1016/j.devcel.2012.10.016.

Notredame, C., D. G. Higgins, and J. Heringa. 2000. "T-Coffee: a novel method for fast and accurate multiple sequence alignment." *J Mol Biol* 302 (1):205–17. doi:10.1006/jmbi.2000.4042.

Ogawa, T., R. Nitta, Y. Okada, and N. Hirokawa. 2004. "A common mechanism for microtubule destabilizers-M type kinesins stabilize curling of the protofilament using the class-specific neck and loops." *Cell* 116 (4):591–602.

Ogawa, T., S. Saijo, N. Shimizu, X. Jiang, and N. Hirokawa. 2017. "Mechanism of catalytic microtubule depolymerization via KIF2-tubulin transitional conformation." *Cell Rep* 20 (11):2626–38. doi:10.1016/j.celrep.2017.08.067.

Peters, C., K. Brejc, L. Belmont, A. J. Bodey, Y. Lee, M. Yu, J. Guo, R. Sakowicz, J. Hartman, and C. A. Moores. 2010. "Insight into the molecular mechanism of the multitasking kinesin-8 motor." *EMBO J* 29 (20):3437–47. doi:10.1038/emboj.2010.220.

Sablin, E. P., F. J. Kull, R. Cooke, R. D. Vale, and R. J. Fletterick. 1996. "Crystal structure of the motor domain of the kinesin-related motor ncd." *Nature* 380 (6574):555–9. doi:10.1038/380555a0.

Sanchez-Perez, Isabel, Steven J. Renwick, Karen Crawley, Inga Karig, Vicky Buck, John C. Meadows, Alejandro Franco-Sanchez, Ursula Fleig, Takashi Toda, and Jonathan B. A. Millar. 2005. "The DASH complex and Klp5/Klp6 kinesin coordinate bipolar chromosome attachment in fission yeast." *EMBO J* 24 (16):2931–43.

Savoian, M. S., M. K. Gatt, M. G. Riparbelli, G. Callaini, and D. M. Glover. 2004. "Drosophila Klp67A is required for proper chromosome congression and segregation during meiosis I." *J Cell Sci* 117 (Pt 16):3669–77. doi:10.1242/jcs.01213.

Savoian, M. S., and D. M. Glover. 2010. "Drosophila Klp67A binds prophase kinetochores to subsequently regulate congression and spindle length." *J Cell Sci* 123 (Pt 5):767–76. doi:10.1242/jcs.055905.

Schiewek, J., U. Schumacher, T. Lange, S. A. Joosse, H. Wikman, K. Pantel, M. Mikhaylova, M. Kneussel, S. Linder, B. Schmalfeldt, L. Oliveira-Ferrer, and S. Windhorst. 2018. "Clinical relevance of cytoskeleton associated proteins for ovarian cancer." *J Cancer Res Clin Oncol* 144 (11):2195–205. doi:10.1007/s00432-018-2710-9.

Shang, Z., K. Zhou, C. Xu, R. Csencsits, J. C. Cochran, and C. V. Sindelar. 2014. "High-resolution structures of kinesin on microtubules provide a basis for nucleotide-gated force-generation." *Elife* 3:e04686. doi:10.7554/eLife.04686.

Shipley, K., M. Hekmat-Nejad, J. Turner, C. Moores, R. Anderson, R. Milligan, R. Sakowicz, and R. Fletterick. 2004. "Structure of a kinesin microtubule depolymerization machine." *EMBO J* 23 (7):1422–32. doi:10.1038/sj.emboj.7600165.

Stout, J. R., A. L. Yount, J. A. Powers, C. Leblanc, S. C. Ems-McClung, and C. E. Walczak. 2011. "Kif18B interacts with EB1 and controls astral microtubule length during mitosis." *Mol Biol Cell* 22 (17):3070–80. doi:10.1091/mbc.E11-04-0363.

Stumpff, J., Y. Du, C. A. English, Z. Maliga, M. Wagenbach, C. L. Asbury, L. Wordeman, and R. Ohi. 2011. "A tethering mechanism controls the processivity and kinetochore-microtubule plus-end enrichment of the kinesin-8 Kif18A." *Mol Cell* 43 (5):764–75. doi:10.1016/j.molcel.2011.07.022.

Stumpff, J., G. von Dassow, M. Wagenbach, C. Asbury, and L. Wordeman. 2008. "The kinesin-8 motor Kif18A suppresses kinetochore movements to control mitotic chromosome alignment." *Dev Cell* 14 (2):252–62. doi:10.1016/j.devcel.2007.11.014.

Su, X., H. Arellano-Santoyo, D. Portran, J. Gaillard, M. Vantard, M. Thery, and D. Pellman. 2013. "Microtubule-sliding activity of a kinesin-8 promotes spindle assembly and spindle-length control." *Nat Cell Biol* 15 (8):948–57. doi:10.1038/ncb2801.

Su, X., W. Qiu, M. L. Gupta, Jr., J. B. Pereira-Leal, S. L. Reck-Peterson, and D. Pellman. 2011. "Mechanisms underlying the dual-mode regulation of microtubule dynamics by Kip3/kinesin-8." *Mol Cell* 43 (5):751–63. doi:10.1016/j.molcel.2011.06.027.

Tan, D., W. J. Rice, and H. Sosa. 2008. "Structure of the kinesin13-microtubule ring complex." *Structure* 16 (11):1732–9. doi:10.1016/j.str.2008.08.017.

Tanenbaum, M. E., L. Macurek, B. van der Vaart, M. Galli, A. Akhmanova, and R. H. Medema. 2011. "A complex of Kif18b and MCAK promotes microtubule depolymerization and is negatively regulated by Aurora kinases." *Curr Biol* 21 (16):1356–65. doi:10.1016/j.cub.2011.07.017.

Unsworth, A., H. Masuda, S. Dhut, and T. Toda. 2008. "Fission yeast kinesin-8 Klp5 and Klp6 are interdependent for mitotic nuclear retention and required for proper microtubule dynamics." *Mol Biol Cell* 19 (12):5104–15.

Varga, V., J. Helenius, K. Tanaka, A. A. Hyman, T. U. Tanaka, and J. Howard. 2006. "Yeast kinesin-8 depolymerizes microtubules in a length-dependent manner." *Nat Cell Biol* 8 (9):957–62. doi:10.1038/ncb1462.

Varga, V., C. Leduc, V. Bormuth, S. Diez, and J. Howard. 2009. "Kinesin-8 motors act cooperatively to mediate length-dependent microtubule depolymerization." *Cell* 138 (6):1174–83. doi:10.1016/j.cell.2009.07.032.

Verhey, K. J., and J. W. Hammond. 2009. "Traffic control: regulation of kinesin motors." *Nat Rev Mol Cell Biol* 10 (11):765–77. doi:10.1038/nrm2782.

Walczak, C. E., H. Zong, S. Jain, and J. R. Stout. 2016. "Spatial regulation of astral microtubule dynamics by Kif18B in PtK cells." *Mol Biol Cell* 27 (20):3021–30. doi:10.1091/mbc.E16-04-0254.

Wang, D., R. Nitta, M. Morikawa, H. Yajima, S. Inoue, H. Shigematsu, M. Kikkawa, and N. Hirokawa. 2016. "Motility and microtubule depolymerization mechanisms of the Kinesin-8 motor, KIF19A." *Elife* 5. doi:10.7554/eLife.18101.

Wang, L., S. Yang, R. Sun, M. Lu, Y. Wu, and Y. Li. 2016. "[Expression of KIF18A in gastric cancer and its association with prognosis]." *Zhonghua Wei Chang Wai Ke Za Zhi* 19 (5):585–9.

Weaver, L. N., S. C. Ems-McClung, J. R. Stout, C. LeBlanc, S. L. Shaw, M. K. Gardner, and C. E. Walczak. 2011. "Kif18A uses a microtubule binding site in the tail for plus-end localization and spindle length regulation." *Curr Biol* 21 (17):1500–6. doi:10.1016/j.cub.2011.08.005.

West, R. R., T. Malmstrom, and J. R. McIntosh. 2001. "Two related kinesins, klp5 (+) and klp6 (+), foster microtubule disassembly and are required for normal chromosome movement in fission yeast." *Mol Biol Cell*, 12:3919–32.

Wickstead, B., K. Gull, and T. A. Richards. 2010. "Patterns of kinesin evolution reveal a complex ancestral eukaryote with a multifunctional cytoskeleton." *BMC Evol Biol* 10:110. doi:10.1186/1471-2148-10-110.

Woehlke, G., A. K. Ruby, C. L. Hart, B. Ly, N. Hom-Booher, and R. D. Vale. 1997. "Microtubule interaction site of the kinesin motor." *Cell* 90 (2):207–16.

Wu, Y., A. Wang, B. Zhu, J. Huang, E. Lu, H. Xu, W. Xia, G. Dong, F. Jiang, and L. Xu. 2018. "KIF18B promotes tumor progression through activating the Wnt/beta-catenin pathway in cervical cancer." *Onco Targets Ther* 11:1707–20. doi:10.2147/OTT.S157440.

Zhang, C., C. Zhu, H. Chen, L. Li, L. Guo, W. Jiang, and S. H. Lu. 2010. "Kif18A is involved in human breast carcinogenesis." *Carcinogenesis* 31 (9):1676–84. doi:10.1093/carcin/bgq134.

Zhu, C., J. Zhao, M. Bibikova, J. D. Leverson, E. Bossy-Wetzel, J. B. Fan, R. T. Abraham, and W. Jiang. 2005. "Functional analysis of human microtubule-based motor proteins, the kinesins and dyneins, in mitosis/cytokinesis using RNA interference." *Mol Biol Cell* 16 (7):3187–99. doi:10.1091/mbc.e05-02-0167.

9 The Kinesin-13 Family
Specialist Destroyers

Julie P. Welburn and Claire T. Friel

CONTENTS

The Kinesin-13 family is a group of specialist microtubule-depolymerising motors. They display no directed motility but disassemble microtubules from either end.

9.1 EXAMPLE FAMILY MEMBERS

Mammalian: KIF2A, KIF2B, KIF2C/MCAK, KIF24
Drosophila melanogaster: KLP10A, KLP59C, KLP59D
Xenopus laevis: XKCM1
Caenorhabditis elegans: KLP7

9.2 STRUCTURAL INFORMATION

Members of this family studied to date function as homodimers. The motor domain is located centrally in the primary sequence and is flanked by divergent N- and C-terminal regions (Figure 9.1A), which are responsible for cellular localisation, regulation and dimerisation. Whilst the Kinesin-13 motor domain alone can depolymerise microtubules (Maney, Wagenbach, and Wordeman 2001, Hertzer et al. 2006, Cooper et al. 2010), the N- and C-terminal flanking regions target the motor domain to specific subcellular structures and are the sites of the majority of regulatory post-translational modifications. Swapping the N termini among the KIF2 Kinesin-13s retargets the motor domains to subcellular regions specified by the N-terminus (Welburn and Cheeseman 2012).

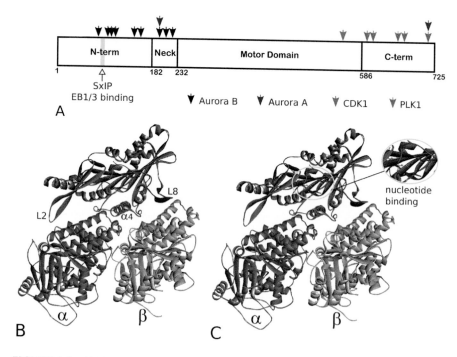

FIGURE 9.1 (A) Typical domain layout for the Kinesin-13 family. Residue numbering is shown according to the sequence of human mitotic centromere-associated kinesin (MCAK). Known phosphorylation sites and regulatory binding sites for human MCAK are highlighted (adapted from Tanenbaum, Medema, and Akhmanova (2011b)). (B), (C) Structure of human MCAK in complex with an α/β-tubulin heterodimer (PDB: 5MIO). (B) Major areas of secondary structure that comprise the tubulin-binding interface: Loop 2 (red), α4-helix (pink) and Loop 8 (blue). (C) Nucleotide-binding site is highlighted inside the yellow oval, plus a magnified view of this region. The major nucleotide-binding motifs are the p-loop (pink), Switch I (blue) and Switch II (red).

The microtubule-binding face of the Kinesin-13 motor domain is composed of Loop 2, the α4 helix, Loop 12, the α5 helix and Loop 8 (Figure 9.1B). Sequence alignments show that the Kinesin-13 motor domain has two family-specific sequence motifs, both of which are found in the tubulin-binding interface. These are (i) the commonly titled 'KVD finger' found in an extended Loop 2 and (ii) a KECIR motif found in the C-terminal half of the α4-helix. Both of these motifs are highly conserved within the Kinesin-13 family and mutations in these motifs can severely impair depolymerisation activity (Ogawa et al. 2004, Shipley et al. 2004, Patel et al. 2016). As with other Kinesin families, the α4-helix forms the central axis of the Kinesin-13 tubulin-binding interface and is found in the intradimer groove between the α- and β-subunits of the same tubulin heterodimer (Wang et al. 2017, Ogawa et al. 2017, Benoit, Asenjo, and Sosa 2018, Trofimova et al. 2018). The Loop 2 region forms a β-hairpin that is extended relative to this region in other kinesin families. The extended Loop 2 is oriented towards the α-subunit end of the heterodimer and contacts the longitudinal interface between β- and α-tubulin known as the 'interdimer groove' (Trofimova et al. 2018, Benoit, Asenjo, and Sosa 2018). Loop 8 contacts the β-tubulin side of the heterodimer and would be oriented toward the plus end of tubulin protofilaments.

The nucleotide-binding site of the Kinesin-13 motor domain is in a similar position to other kinesin families and contains the same four conserved ATP-binding motifs: the p-loop (N1), Switch I (N2) and Switch II (N3) and RxRP (N4). These motifs appear to play similar roles in binding and hydrolysis of the nucleotide and in communicating the nucleotide state of the motor domain to the microtubule-binding face, as in other kinesins. However, the ATP turnover cycle of the Kinesin-13 MCAK (mitotic centromere-associated kinesin) is atypical among the Kinesin superfamily. In the absence of tubulin, ATP cleavage rather than ADP dissociation is rate-limiting and only tubulin at or near the microtubule end accelerates ADP dissociation from the MCAK motor domain (Friel and Howard 2011, Hunter et al. 2003). This atypical ATP turnover cycle is likely to be conserved among the Kinesin-13 family and adapts members of this family to their microtubule-depolymerising function.

A short, typically positively charged region, called the neck, lies N-terminal to the motor domain (Figure 9.1A). This neck region does not appear to play the same role as the region of this name found in translocating kinesins. For MCAK, the neck region is required for full depolymerisation activity. Removal or neutralisation of the positively charged neck markedly reduces MCAK's depolymerisation activity, and studies suggest that the neck enhances delivery of MCAK to the microtubule ends (Ovechkina, Wagenbach, and Wordeman 2002, Cooper et al. 2010). Work from multiple labs reveals that the neck and N-terminal regions of Kinesin-13s associate with the tubulin heterodimer adjacent to that occupied by the motor domain. The cryo-electron microscopy (cryo-EM) structure of Klp10A on microtubules indicates that the neck linker occupies the tubulin heterodimer prior to the one occupied by the motor domain, creating steric hindrance to a second motor domain (Benoit, Asenjo, and Sosa 2018). Both negative-stain electron microscopy and crosslinking studies analysed by mass spectrometry indicate that the structure of both KIF2A and MCAK is compact in solution, with the N- and C-terminal regions interacting with each other and with the motor domain (Noda et al. 1995, Maney, Wagenbach, and Wordeman 2001, McHugh et al. 2019). Upon interaction with the microtubule, MCAK has been shown to extend, allowing the neck and N-terminal region to interact with the tubulin heterodimer longitudinally adjacent to that to which the motor domain is bound (McHugh et al. 2019). It is still not known where the second motor domain and the C-termini are positioned when the first motor and the N-terminus occupy two longitudinally adjacent tubulin heterodimers.

9.3 FUNCTIONAL PROPERTIES

The Kinesin-13 family are a group of microtubule-depolymerising kinesins and members of this family are suggested to be major regulators of microtubule length. The most highly studied Kinesin-13, MCAK, has no translocating activity but displays diffusive movement on the microtubule lattice with no directional bias (Figure 9.2A) (Helenius et al. 2006). MCAK binds tightly to the ends of microtubules and removes tubulin dimers, causing depolymerisation from both the plus and

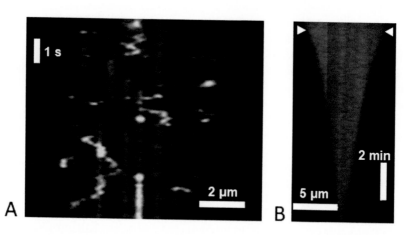

FIGURE 9.2   The kinesin-13, mitotic centromere-associated kinesin (MCAK), diffuses on the microtubule lattice and depolymerises microtubules from both the plus and minus ends. (A) Kymograph showing the diffusive movement of green fluorescent protein (GFP)–MCAK on a microtubule. (B) Kymograph showing the depolymerisation of a microtubule by MCAK. Addition of MCAK is shown by the arrowheads.

minus ends of the microtubule at a similar rate (Figure 9.2B) (Helenius et al. 2006). This microtubule-depolymerising function appears to be characteristic of members of the Kinesin-13 family. KIF2A and MCAK both show potent microtubule depolymerase activity in cells and *in vitro* (Knowlton et al. 2009, Lan et al. 2004, Maney, Wagenbach, and Wordeman 2001, Hunter et al. 2003). KIF24 has been shown to depolymerise microtubules *in vitro* (Kobayashi et al. 2011). The *Xenopus* homologue of MCAK, XKCM1, is a microtubule depolymeriser (Walczak, Mitchison, and Desai 1996, Desai et al. 1999) and both the *Drosophila* and *C. elegans* Kinesin-13s, KLP10A and KLP7, exhibit microtubule-destabilising activity in cells (Gigant et al. 2017, Goshima and Vale 2003).

Unlike the majority of other kinesins, ATP turnover by Kinesin-13s is not significantly accelerated by the microtubule lattice. Motor activity for members of this family occurs at the ends of microtubules and, accordingly, ATP turnover is significantly accelerated only by microtubule ends (Hunter et al. 2003, Friel and Howard 2011). The Kinesin-13, MCAK, has been shown to have an atypical ATP turnover cycle in that the rate-limiting step in the absence of tubulin is ATP cleavage rather than ADP dissociation (Friel and Howard 2011). This atypical ATP turnover cycle underlies MCAK's behaviour as a specialist depolymerising kinesin. In solution, MCAK is predominantly ATP-bound and meets the microtubule in this nucleotide state. This contrasts with all translocating kinesins studied to date, which meet the microtubule in an ADP-bound state (Hackney 1988). ATP cleavage is stimulated by interaction with tubulin and so MCAK is driven into the ADP-bound state by interaction with microtubules. ADP.MCAK forms a non-specific diffusive interaction with the microtubule lattice (Helenius et al. 2006). Only the microtubule end accelerates dissociation of ADP, driving MCAK into an ATP-bound depolymerisation competent complex with tubulin (Friel and Howard 2011).

Kinesin-13s likely drive microtubule depolymerisation by stabilising a curved conformation of tubulin that destabilises the microtubule. Kinesin-13s have been shown to stabilise microtubule protofilament curls and rings in the presence of a non-hydrolysable ATP analogue (Tan et al. 2006). The ability of the ATP-bound Kinesin-13 motor domain to stabilise tubulin heterodimers in a curved conformation (Moores and Milligan 2008, Tan et al. 2006), coupled with an atypical ATP turnover cycle which promotes formation of a depolymerisation-competent complex with tubulin specifically at the microtubule end (Friel and Howard 2011), results in a kinesin motor domain specialised for microtubule depolymerisation.

9.4 PHYSIOLOGICAL ROLES

Kinesin-13s play prominent roles in regulating microtubule length. Members of this family are particularly important during meiosis and mitosis in higher eukaryotes. There are four mammalian Kinesin13s: KIF2A, KIF2B, MCAK/KIF2C and KIF24. The KIF2s are closely related and part of the Kinesin-13B subfamily, whereas KIF24 is a member of the larger Kinesin-13C subfamily (Wickstead, Gull, and Richards 2010). These and Kinesin-13s from other species play important roles in multiple processes ranging from spindle assembly and chromosome segregation to cilium length control and neuronal development and differentiation (Manning et al. 2007, Welburn and Cheeseman 2012, Homma et al. 2003, Kobayashi et al. 2011, Domnitz et al. 2012, Maor-Nof et al. 2013).

The most highly studied member of the Kinesin-13 family, MCAK, was discovered as a kinesin associated with the centromere of mitotic chromosomes and named Mitotic Centromere-Associated Kinesin (Wordeman and Mitchison 1995). MCAK is recruited to the centromere at prophase and remains centromere associated until after telophase (Maney et al. 1998). MCAK is also present, although at lower levels, at the growing plus ends of microtubules, via an association with End-Binding (EB) proteins (Moore et al. 2005, Lee et al. 2008, Honnappa et al. 2009).

Kinesin-13s are important in regulating spindle assembly and chromosome segregation in both mitotic and meiotic cell division. KIF2A regulates the length of mitotic spindles formed in *Xenopus* extracts (Wilbur and Heald 2013), and in cultured human cells it appears to have a significant role in regulating the length of the central spindle region (Uehara et al. 2013). Some studies show KIF2A to be required for the formation of a bipolar spindle (Manning et al. 2007, Ganem and Compton 2004). Knockdown of KIF2A in mouse oocytes leads to defects in meiotic spindle formation (Yi et al. 2016, Chen et al. 2016). KIF2C/MCAK localises to mitotic spindle poles, spindle midzone and kinetochores (Wordeman and Mitchison 1995). Disruption of MCAK function has little effect on spindle assembly but results in an increase in the frequency of lagging chromosomes during anaphase (Ganem, Upton, and Compton 2005, Maney et al. 1998, Lan et al. 2004). Specific depletion of centromere-associated MCAK decreases the directional coordination between sister kinetochores and inhibits efficient detaching of centromeres from kinetochore microtubules (Wordeman, Wagenbach, and von Dassow 2007). MCAK likely has a role in the correction of kinetochore–microtubule attachment errors.

In addition to its spindle localisation, MCAK is found at the tips of growing microtubules (Moore et al. 2005). MCAK possesses an SxIP motif which mediates association with EB proteins, facilitating microtubule plus-end targeting (Honnappa et al. 2009). MCAK also relies on an interaction with the Kinesin-8 KIF18B for plus-end localisation (Tanenbaum et al. 2011a). At microtubule plus ends, MCAK regulates microtubule length in interphase and mitosis. Astral microtubule length control during mitosis is particularly important to balance forces in the spindle and ensure spindle centring (Rankin and Wordeman 2010, McHugh, Gluszek, and Welburn 2018). In human cells, excessive length of astral microtubules can cause the spindle to collapse (van Heesbeen et al. 2017).

The role of KIF2B remains controversial, as the phenotype implicating it in kinetochore–microtubule attachment could not be rescued or reproduced by other researchers in the field (Bakhoum et al. 2009). Studies of depletion and corresponding rescue of KIF2B have not yet led to any conclusive phenotype (Welburn and Cheeseman 2012, Tanenbaum et al. 2011a).

Studies in *Xenopus* extracts have shown that displacement of the Kinesin-13 XKCM1 from centromeres leads to chromosome misalignment (Walczak et al. 2002), suggesting that its function, similar to that proposed for MCAK, is to enhance depolymerisation of microtubules that are not properly attached to kinetochores. In *Drosophila* the three identified Kinesin-13s all contribute to spindle microtubule dynamics (Mennella et al. 2005, Rogers et al. 2004, Goshima and Vale 2003). KLP10A targets spindle poles and depolymerises microtubules at their minus end, contributing to chromosome movement by means of poleward flux. By contrast, KLP59C exerts its activity at centromeres, depolymerising spindle microtubules at their plus ends (Rogers et al. 2004). The third *Drosophila* Kinesin-13, Klp59D, appears to depolymerise microtubules from both ends (Rath et al. 2009).

In addition to spindle-specific activity, Kinesin-13s have also been shown to promote destabilisation of the whole microtubule network. This activity appears to be necessary to counteract ectopic, spontaneous microtubule assembly to enable correct spindle assembly in the absence of centrosomes. Microtubule-destabilising activity of the *C. elegans* Kinesin-13, KLP7, is required during meiotic spindle organisation which occurs without centrosomes (Gigant et al. 2017). This global microtubule destabilisation activity seems to be a general function of Kinesin-13s. KLP-7 and MCAK have both been shown to prevent incorrect microtubule assembly in mitotic cells when centrosome activity is reduced or absent (Gigant et al. 2017, Srayko et al. 2005, Garzon-Coral, Fantana, and Howard 2016).

KIF2A plays an important role in neuronal development. Mice lacking KIF2A die within a day of birth, displaying multiple brain abnormalities (Homma et al. 2003). Inducible knockdown of KIF2A shows that the activity of its protein product is also crucial to brain development after birth (Homma et al. 2018). The major phenotype connected with lack of KIF2A activity, both pre and postnatal, is abnormally long axonal collateral branches. KIF2A activity is suggested to suppress the growth of aberrant axons by suppressing microtubule polymerisation at the growth cone.

The centrosome-localised Kinesin-13s, KIF24 and KIF2A, are implicated in regulating cilium assembly/disassembly (Kobayashi et al. 2011, Miyamoto et al.

2015, Kim et al. 2015). This role is regulated by phosphorylation for both these Kinesin-13s. Phosphorylation of KIF24 by Nek2 (Kim et al. 2015) and of KIF2A by PLK1 (Miyamoto et al. 2015) enhances microtubule-depolymerisation activity and promotes cilium disassembly. In addition to its role in regulating dynamics of cytoplasmic and spindle microtubules, the *Drosophila* Kinesin-13, KLP10A, regulates centriole length (Delgehyr et al. 2012). It is possible that there is a competition for centriole ends between the centriole-capping protein CP110 and the depolymerising Kinesin-13 to control centriole, primary cilium and flagellum length (Delgehyr et al. 2012, Kim et al. 2015).

The Kinesin-13 family of microtubule depolymerases play essential roles throughout Eukaryotes as major regulators of microtubule length and in the regulation of cellular architecture and function.

9.4.1 KINESIN-13 REGULATION

Kinesin-13s are potent microtubule depolymerases and so their activity is typically tightly regulated, particularly during cell division. Members of the Kinesin-13 family studied to date are regulated either through phosphorylation or via interaction with other proteins. Regulatory modifications or interactions can direct the subcellular localisation of the Kinesin-13 and/or alter depolymerising activity, either directly or by affecting protein stability.

The Aurora kinases phosphorylate MCAK at multiple residues in the N-terminal region and one residue in the neck region (Ser192 in humans, Ser196 in *Xenopus*) (Andrews et al. 2004, Lan et al. 2004, Ohi et al. 2004, Zhang, Ems-McClung, and Walczak 2008, Ritter et al. 2015). Phosphorylation by Aurora B both affects the subcellular localisation of MCAK and directly inhibits its depolymerisation activity. Aurora B phosphorylation is required to localise MCAK to centromeres (Ohi et al. 2004, Lan et al. 2004). Inhibition of depolymerisation activity as a result of phosphorylation by Aurora B has been shown to occur via a phosphorylation-dependent conformational change that reduces MCAKs affinity for microtubules (Ems-McClung et al. 2013). Inhibition of activity occurs in a graded fashion, dependent on the number of sites phosphorylated (McHugh et al. 2019).

MCAK is phosphorylated within the motor domain by cyclin-dependant kinase 1 (CDK1); both the localisation and the activity of MCAK are regulated by phosphorylation at this site (Sanhaji et al. 2010). CDK1 phosphorylation promotes the release of MCAK from centromeres and also inhibits microtubule depolymerisation activity by directly disrupting the ability of MCAK to recognise the microtubule end (Belsham and Friel 2017). The microtubule-depolymerising activity of the *Drosophila* Kinesin-13, KLP10A, is also reduced by phosphorylation in the motor domain, possibly by a similar mechanism (Mennella et al. 2009).

MCAK is phosphorylated by polo-like kinase 1 (PLK1) at several possible sites in the C-terminal region (Zhang et al. 2011). Of these, S715 appears most critical, and phosphorylation at this site promotes depolymerisation activity (Shao et al. 2015). Phosphorylation by PLK1 and the Aurora kinases at sites in the C-terminal region affect MCAK's depolymerisation activity and localisation via conformational changes driven by disruption of the interaction of the C-terminus with the neck and/

or motor domain (Zhang, Ems-McClung, and Walczak 2008, Ems-McClung et al. 2013, Zong et al. 2016, Talapatra, Harker, and Welburn 2015).

Kinesin-13s are also regulated via interaction with other proteins and these interactions are often themselves regulated by phosphorylation. The interaction of MCAK with EB proteins that mediate microtubule plus tip tracking is regulated by phosphorylation. Sites in the N-terminal region of MCAK phosphorylated by Aurora B are close to the SxIP motif recognised by EB proteins, and phosphorylation at these sites disrupts the EB–MCAK interaction through electrostatic interactions and abolishes microtubule tip tracking by MCAK in cells (Honnappa et al. 2009, Moore et al. 2005). The centromere-targeting of MCAK is dependent on interaction with the Shugoshin family member hSgo2 (Huang et al. 2007). The interaction between hSgo2 and MCAK requires phosphorylation of hSgo2 by Aurora B kinase (Tanno et al. 2010).

Other proteins also regulate Kinesin-13 activity, such as Tip150 and GTSE1, although the mechanisms of regulation, whether direct or indirect, remain unclear (Bendre et al. 2016, Jiang et al. 2009). Although much is known about how Kinesin-13 activity is regulated in cells, the regulation networks are complex and further work is required to fully understand the regulatory mechanisms that govern Kinesin-13 function.

9.5   INVOLVEMENT IN DISEASE

In accordance with the role of the Kinesin-13s in spindle formation and faithful chromosome separation during cell division, dysregulated Kinesin-13 expression is found in many types of cancer tissue. Elevated expression of MCAK is observed in breast, colorectal and gastric cancers (Shimo et al. 2008, Nakamura et al. 2007) and the available data suggests that enhanced MCAK levels are associated with increased invasiveness and metastasis, poor survival rates and resistance to anticancer therapies (Ishikawa et al. 2008, Ganguly, Yang, and Cabral 2011, Ganguly et al. 2011). High KIF2A expression is suggested to promote proliferation, migration and to predict poor prognosis in lung cancer (Xie et al. 2018, Bai et al. 2019) and to be associated with poor prognosis in ovarian cancer and lymphoma (Zhang et al. 2017, Wang et al. 2016, Zhang et al. 2016, Sheng et al. 2018).

KIF2A plays a major role in neuronal development, proliferation and organisation. Correspondingly, mutations in KIF2A are reported in patients with neuronal development disorders such as lissencephaly, microcephaly and pachygyria (Cavallin et al. 2017, Tian et al. 2016, Poirier et al. 2013).

REFERENCES

Andrews, P. D., Y. Ovechkina, N. Morrice, M. Wagenbach, K. Duncan, L. Wordeman, and J. R. Swedlow. 2004. "Aurora B regulates MCAK at the mitotic centromere." *Dev Cell* 6 (2):253–68.

Bai, Y., L. Xiong, M. Zhu, Z. Yang, J. Zhao, and H. Tang. 2019. "Co-expression network analysis identified KIF2C in association with progression and prognosis in lung adeno-carcinoma." *Cancer Biomark* 24 (3):371–82. doi:10.3233/CBM-181512.

Bakhoum, S. F., S. L. Thompson, A. L. Manning, and D. A. Compton. 2009. "Genome stability is ensured by temporal control of kinetochore-microtubule dynamics." *Nat Cell Biol* 11 (1):27–35. doi:ncb1809 [pii] 10.1038/ncb1809.

Belsham, H. R., and C. T. Friel. 2017. "A Cdk1 phosphomimic mutant of MCAK impairs microtubule end recognition." *PeerJ* 5:e4034. doi:10.7717/peerj.4034.

Bendre, S., A. Rondelet, C. Hall, N. Schmidt, Y. C. Lin, G. J. Brouhard, and A. W. Bird. 2016. "GTSE1 tunes microtubule stability for chromosome alignment and segregation by inhibiting the microtubule depolymerase MCAK." *J Cell Biol* 215 (5):631–47. doi:10.1083/jcb.201606081.

Benoit, Mpmh, A. B. Asenjo, and H. Sosa. 2018. "Cryo-EM reveals the structural basis of microtubule depolymerization by kinesin-13s." *Nat Commun* 9 (1):1662. doi:10.1038/s41467-018-04044-8.

Cavallin, M., E. K. Bijlsma, A. El Morjani, S. Moutton, E. A. Peeters, C. Maillard, J. M. Pedespan, A. M. Guerrot, V. Drouin-Garaud, C. Coubes, D. Genevieve, C. Bole-Feysot, C. Fourrage, J. Steffann, and N. Bahi-Buisson. 2017. "Recurrent KIF2A mutations are responsible for classic lissencephaly." *Neurogenetics* 18 (2):73–9. doi:10.1007/s10048-016-0499-8.

Chen, M. H., Y. Liu, Y. L. Wang, R. Liu, B. H. Xu, F. Zhang, F. P. Li, X. Lu, Y. H. Lin, S. W. He, B. Q. Liao, X. P. Fu, X. X. Wang, X. J. Yang, and H. L. Wang. 2016. "KIF2A regulates the spindle assembly and the metaphase I-anaphase I transition in mouse oocyte." *Sci Rep* 6:39337. doi:10.1038/srep39337.

Cooper, J. R., M. Wagenbach, C. L. Asbury, and L. Wordeman. 2010. "Catalysis of the microtubule on-rate is the major parameter regulating the depolymerase activity of MCAK." *Nat Struct Mol Biol* 17 (1):77–82. doi:10.1038/nsmb.1728.

Delgehyr, N., H. Rangone, J. Fu, G. Mao, B. Tom, M. G. Riparbelli, G. Callaini, and D. M. Glover. 2012. "Klp10A, a microtubule-depolymerizing kinesin-13, cooperates with CP110 to control Drosophila centriole length." *Curr Biol* 22 (6):502–9. doi:10.1016/j.cub.2012.01.046.

Desai, A., S. Verma, T. J. Mitchison, and C. E. Walczak. 1999. "Kin I kinesins are microtubule-destabilizing enzymes." *Cell* 96 (1):69–78.

Domnitz, S. B., M. Wagenbach, J. Decarreau, and L. Wordeman. 2012. "MCAK activity at microtubule tips regulates spindle microtubule length to promote robust kinetochore attachment." *J Cell Biol* 197 (2):231–7. doi:jcb.201108147 [pii] 10.1083/jcb.201108147.

Ems-McClung, S. C., S. G. Hainline, J. Devare, H. Zong, S. Cai, S. K. Carnes, S. L. Shaw, and C. E. Walczak. 2013. "Aurora B inhibits MCAK activity through a phosphoconformational switch that reduces microtubule association." *Curr Biol* 23 (24):2491–9. doi:10.1016/j.cub.2013.10.054.

Friel, C. T., and J. Howard. 2011. "The kinesin-13 MCAK has an unconventional ATPase cycle adapted for microtubule depolymerization." *EMBO J* 30 (19):3928–39. doi:emboj2011290 [pii] 10.1038/emboj.2011.290.

Ganem, N. J., and D. A. Compton. 2004. "The KinI kinesin Kif2a is required for bipolar spindle assembly through a functional relationship with MCAK." *J Cell Biol* 166 (4):473–8. doi:10.1083/jcb.200404012jcb.200404012 [pii].

Ganem, N. J., K. Upton, and D. A. Compton. 2005. "Efficient mitosis in human cells lacking poleward microtubule flux." *Curr Biol* 15 (20):1827–32. doi:S0960-9822(05)01035-3 [pii] 10.1016/j.cub.2005.08.065.

Ganguly, A., H. Yang, and F. Cabral. 2011. "Overexpression of mitotic centromere-associated Kinesin stimulates microtubule detachment and confers resistance to paclitaxel." *Mol Cancer Ther* 10 (6):929–37. doi:10.1158/1535-7163.MCT-10-1109.

Ganguly, A., H. Yang, M. Pedroza, R. Bhattacharya, and F. Cabral. 2011. "Mitotic centromere-associated kinesin (MCAK) mediates paclitaxel resistance." *J Biol Chem* 286 (42):36378–84. doi:10.1074/jbc.M111.296483.

Garzon-Coral, C., H. A. Fantana, and J. Howard. 2016. "A force-generating machinery maintains the spindle at the cell center during mitosis." *Science* 352 (6289):1124–7. doi:10.1126/science.aad9745.

Gigant, E., M. Stefanutti, K. Laband, A. Gluszek-Kustusz, F. Edwards, B. Lacroix, G. Maton, J. C. Canman, J. P. I. Welburn, and J. Dumont. 2017. "Inhibition of ectopic microtubule assembly by the kinesin-13 KLP-7 prevents chromosome segregation and cytokinesis defects in oocytes." *Development* 144 (9):1674–86. doi:10.1242/dev.147504.

Goshima, G., and R. D. Vale. 2003. "The roles of microtubule-based motor proteins in mitosis: comprehensive RNAi analysis in the Drosophila S2 cell line." *J Cell Biol* 162 (6):1003–16. doi:10.1083/jcb.200303022.

Hackney, D. D. 1988. "Kinesin ATPase: rate-limiting ADP release." *Proc Natl Acad Sci U S A* 85 (17):6314–8.

Helenius, J., G. Brouhard, Y. Kalaidzidis, S. Diez, and J. Howard. 2006. "The depolymerizing kinesin MCAK uses lattice diffusion to rapidly target microtubule ends." *Nature* 441:115–119. doi.org/10.1038/nature04736.

Hertzer, K. M., S. C. Ems-McClung, S. L. Kline-Smith, T. G. Lipkin, S. P. Gilbert, and C. E. Walczak. 2006. "Full-length dimeric MCAK is a more efficient microtubule depolymerase than minimal domain monomeric MCAK." *Mol Biol Cell* 17 (2):700–10. doi:E05-08-0821 [pii] 10.1091/mbc.E05-08-0821.

Homma, N., Y. Takei, Y. Tanaka, T. Nakata, S. Terada, M. Kikkawa, Y. Noda, and N. Hirokawa. 2003. "Kinesin superfamily protein 2A (KIF2A) functions in suppression of collateral branch extension." *Cell* 114 (2):229–39.

Homma, N., R. Zhou, M. I. Naseer, A. G. Chaudhary, M. H. Al-Qahtani, and N. Hirokawa. 2018. "KIF2A regulates the development of dentate granule cells and postnatal hippocampal wiring." *Elife* 7. doi:10.7554/eLife.30935.

Honnappa, S., S. M. Gouveia, A. Weisbrich, F. F. Damberger, N. S. Bhavesh, H. Jawhari, I. Grigoriev, F. J. van Rijssel, R. M. Buey, A. Lawera, I. Jelesarov, F. K. Winkler, K. Wuthrich, A. Akhmanova, and M. O. Steinmetz. 2009. "An EB1-binding motif acts as a microtubule tip localization signal." *Cell* 138 (2):366–76. doi:S0092-8674(09)00638-2 [pii] 10.1016/j.cell.2009.04.065.

Huang, H., J. Feng, J. Famulski, J. B. Rattner, S. T. Liu, G. D. Kao, R. Muschel, G. K. Chan, and T. J. Yen. 2007. "Tripin/hSgo2 recruits MCAK to the inner centromere to correct defective kinetochore attachments." *J Cell Biol* 177 (3):413–24. doi:jcb.200701122 [pii] 10.1083/jcb.200701122.

Hunter, A. W., M. Caplow, D. L. Coy, W. O. Hancock, S. Diez, L. Wordeman, and J. Howard. 2003. "The kinesin-related protein MCAK is a microtubule depolymerase that forms an ATP-hydrolyzing complex at microtubule ends." *Mol Cell* 11 (2):445–57. doi:S1097276503000492 [pii].

Ishikawa, K., Y. Kamohara, F. Tanaka, N. Haraguchi, K. Mimori, H. Inoue, and M. Mori. 2008. "Mitotic centromere-associated kinesin is a novel marker for prognosis and lymph node metastasis in colorectal cancer." *Br J Cancer* 98 (11):1824–9. doi:10.1038/sj.bjc.6604379.

Jiang, K., J. Wang, J. Liu, T. Ward, L. Wordeman, A. Davidson, F. Wang, and X. Yao. 2009. "TIP150 interacts with and targets MCAK at the microtubule plus ends." *EMBO Rep* 10 (8):857–65. doi:10.1038/embor.2009.94.

Kim, S., K. Lee, J. H. Choi, N. Ringstad, and B. D. Dynlacht. 2015. "Nek2 activation of Kif24 ensures cilium disassembly during the cell cycle." *Nat Commun* 6:8087. doi:10.1038/ncomms9087.

Knowlton, A. L., V. V. Vorozhko, W. Lan, G. J. Gorbsky, and P. T. Stukenberg. 2009. "ICIS and Aurora B coregulate the microtubule depolymerase Kif2a." *Curr Biol* 19 (9):758–63. doi:10.1016/j.cub.2009.03.018.

Kobayashi, T., W. Y. Tsang, J. Li, W. Lane, and B. D. Dynlacht. 2011. "Centriolar kinesin Kif24 interacts with CP110 to remodel microtubules and regulate ciliogenesis." *Cell* 145 (6):914–25. doi:10.1016/j.cell.2011.04.028.

Lan, W., X. Zhang, S. L. Kline-Smith, S. E. Rosasco, G. A. Barrett-Wilt, J. Shabanowitz, D. F. Hunt, C. E. Walczak, and P. T. Stukenberg. 2004. "Aurora B phosphorylates centromeric MCAK and regulates its localization and microtubule depolymerization activity." *Curr Biol* 14 (4):273–86. doi:10.1016/j.cub.2004.01.055.

Lee, T., K. J. Langford, J. M. Askham, A. Bruning-Richardson, and E. E. Morrison. 2008. "MCAK associates with EB1." *Oncogene* 27 (17):2494–500. doi:10.1038/sj.onc.1210867.

Maney, T., A. W. Hunter, M. Wagenbach, and L. Wordeman. 1998. "Mitotic centromere-associated kinesin is important for anaphase chromosome segregation." *J Cell Biol* 142 (3):787–801.

Maney, T., M. Wagenbach, and L. Wordeman. 2001. "Molecular dissection of the microtubule depolymerizing activity of mitotic centromere-associated kinesin." *J Biol Chem* 276 (37):34753–8. doi:10.1074/jbc.M106626200

Manning, A. L., N. J. Ganem, S. F. Bakhoum, M. Wagenbach, L. Wordeman, and D. A. Compton. 2007. "The kinesin-13 proteins Kif2a, Kif2b, and Kif2c/MCAK have distinct roles during mitosis in human cells." *Mol Biol Cell* 18 (8):2970–9. doi:10.1091/mbc.e07-02-0110.

Maor-Nof, M., N. Homma, C. Raanan, A. Nof, N. Hirokawa, and A. Yaron. 2013. "Axonal pruning is actively regulated by the microtubule-destabilizing protein kinesin superfamily protein 2A." *Cell Rep* 3 (4):971–7. doi:10.1016/j.celrep.2013.03.005.

McHugh, T., A. A. Gluszek, and J. P. I. Welburn. 2018. "Microtubule end tethering of a processive kinesin-8 motor Kif18b is required for spindle positioning." *J Cell Biol* 217 (7):2403–16. doi:10.1083/jcb.201705209.

McHugh, T., J. Zou, V. A. Volkov, A. Bertin, S. K. Talapatra, J. Rappsilber, M. Dogterom, and J. P. I. Welburn. 2019. "The depolymerase activity of MCAK shows a graded response to Aurora B kinase phosphorylation through allosteric regulation." *J Cell Sci* 132 (4). doi:10.1242/jcs.228353.

Mennella, V., G. C. Rogers, S. L. Rogers, D. W. Buster, R. D. Vale, and D. J. Sharp. 2005. "Functionally distinct kinesin-13 family members cooperate to regulate microtubule dynamics during interphase." *Nat Cell Biol* 7 (3):235–45. doi:10.1038/ncb1222.

Mennella, V., D. Y. Tan, D. W. Buster, A. B. Asenjo, U. Rath, A. Ma, H. J. Sosa, and D. J. Sharp. 2009. "Motor domain phosphorylation and regulation of the Drosophila kinesin 13, KLP10A." *J Cell Biol* 186 (4):481–90. doi:10.1083/jcb.200902113.

Miyamoto, T., K. Hosoba, H. Ochiai, E. Royba, H. Izumi, T. Sakuma, T. Yamamoto, B. D. Dynlacht, and S. Matsuura. 2015. "The microtubule-depolymerizing activity of a mitotic kinesin protein KIF2A drives primary cilia disassembly coupled with cell proliferation." *Cell Rep* 10 (5):664–73. doi:10.1016/j.celrep.2015.01.003.

Moore, A. T., K. E. Rankin, G. von Dassow, L. Peris, M. Wagenbach, Y. Ovechkina, A. Andrieux, D. Job, and L. Wordeman. 2005. "MCAK associates with the tips of polymerizing microtubules." *J Cell Biol* 169 (3):391–7. doi:jcb.200411089 [pii] 10.1083/jcb.200411089.

Moores, C. A., and R. A. Milligan. 2008. "Visualisation of a kinesin-13 motor on microtubule end mimics." *J Mol Biol* 377 (3):647–54. doi:S0022-2836(08)00135-6 [pii] 10.1016/j.jmb.2008.01.079.

Nakamura, Y., F. Tanaka, N. Haraguchi, K. Mimori, T. Matsumoto, H. Inoue, K. Yanaga, and M. Mori. 2007. "Clinicopathological and biological significance of mitotic centromere-associated kinesin overexpression in human gastric cancer." *Br J Cancer* 97 (4):543–9. doi:10.1038/sj.bjc.6603905.

Noda, Y., R. Sato-Yoshitake, S. Kondo, M. Nangaku, and N. Hirokawa. 1995. "KIF2 is a new microtubule-based anterograde motor that transports membranous organelles distinct from those carried by kinesin heavy chain or KIF3A/B." *J Cell Biol* 129 (1):157–67. doi:10.1083/jcb.129.1.157.

Ogawa, T., R. Nitta, Y. Okada, and N. Hirokawa. 2004. "A common mechanism for microtubule destabilizers-M type kinesins stabilize curling of the protofilament using the class-specific neck and loops." *Cell* 116 (4):591–602. doi:S0092867404001291 [pii].

Ogawa, T., S. Saijo, N. Shimizu, X. Jiang, and N. Hirokawa. 2017. "Mechanism of catalytic microtubule depolymerization via KIF2-tubulin transitional conformation." *Cell Rep* 20 (11):2626–38. doi:10.1016/j.celrep.2017.08.067.

Ohi, R., T. Sapra, J. Howard, and T. J. Mitchison. 2004. "Differentiation of cytoplasmic and meiotic spindle assembly MCAK functions by Aurora B-dependent phosphorylation." *Mol Biol Cell* 15 (6):2895–906. doi:10.1091/mbc.e04-02-0082.

Ovechkina, Y., M. Wagenbach, and L. Wordeman. 2002. "K-loop insertion restores microtubule depolymerizing activity of a 'neckless' MCAK mutant." *J Cell Biol* 159 (4):557–62. doi:10.1083/jcb.200205089.

Patel, J. T., H. R. Belsham, A. J. Rathbone, B. Wickstead, C. Gell, and C. T. Friel. 2016. "The family-specific alpha4-helix of the kinesin-13, MCAK, is critical to microtubule end recognition." *Open Biol* 6 (10). doi:10.1098/rsob.160223.

Poirier, K., N. Lebrun, L. Broix, G. Tian, Y. Saillour, C. Boscheron, E. Parrini, S. Valence, B. S. Pierre, M. Oger, D. Lacombe, D. Genevieve, E. Fontana, F. Darra, C. Cances, M. Barth, D. Bonneau, B. D. Bernadina, S. N'Guyen, C. Gitiaux, P. Parent, V. des Portes, J. M. Pedespan, V. Legrez, L. Castelnau-Ptakine, P. Nitschke, T. Hieu, C. Masson, D. Zelenika, A. Andrieux, F. Francis, R. Guerrini, N. J. Cowan, N. Bahi-Buisson, and J. Chelly. 2013. "Mutations in TUBG1, DYNC1H1, KIF5C and KIF2A cause malformations of cortical development and microcephaly." *Nat Genet* 45 (6):639–47. doi:10.1038/ng.2613.

Rankin, K. E., and L. Wordeman. 2010. "Long astral microtubules uncouple mitotic spindles from the cytokinetic furrow." *J Cell Biol* 190 (1):35–43. doi:10.1083/jcb.201004017.

Rath, U., G. C. Rogers, D. Tan, M. A. Gomez-Ferreria, D. W. Buster, H. J. Sosa, and D. J. Sharp. 2009. "The Drosophila kinesin-13, KLP59D, impacts Pacman- and Flux-based chromosome movement." *Mol Biol Cell* 20 (22):4696–705. doi:E09-07-0557 [pii] 10.1091/mbc.E09-07-0557.

Ritter, A., M. Sanhaji, A. Friemel, S. Roth, U. Rolle, F. Louwen, and J. Yuan. 2015. "Functional analysis of phosphorylation of the mitotic centromere-associated kinesin by Aurora B kinase in human tumor cells." *Cell Cycle* 14 (23):3755–67. doi:10.1080/15384101.2015.1068481.

Rogers, G. C., S. L. Rogers, T. A. Schwimmer, S. C. Ems-McClung, C. E. Walczak, R. D. Vale, J. M. Scholey, and D. J. Sharp. 2004. "Two mitotic kinesins cooperate to drive sister chromatid separation during anaphase." *Nature* 427 (6972):364–70. doi:10.1038/nature02256.

Sanhaji, M., C. T. Friel, N. N. Kreis, A. Kramer, C. Martin, J. Howard, K. Strebhardt, and J. Yuan. 2010. "Functional and spatial regulation of mitotic centromere-associated kinesin by cyclin-dependent kinase 1." *Mol Cell Biol* 30 (11):2594–607. doi:MCB.00098-10 [pii] 10.1128/MCB.00098-10.

Shao, H., Y. Huang, L. Zhang, K. Yuan, Y. Chu, Z. Dou, C. Jin, M. Garcia-Barrio, X. Liu, and X. Yao. 2015. "Spatiotemporal dynamics of Aurora B-PLK1-MCAK signaling axis orchestrates kinetochore bi-orientation and faithful chromosome segregation." *Sci Rep* 5:12204. doi:10.1038/srep12204.

Sheng, N., Y. Z. Xu, Q. H. Xi, H. Y. Jiang, C. Y. Wang, Y. Zhang, and Q. Ye. 2018. "Overexpression of KIF2A is suppressed by miR-206 and associated with poor prognosis in ovarian cancer." *Cell Physiol Biochem* 50 (3):810–22. doi:10.1159/000494467.

Shimo, A., C. Tanikawa, T. Nishidate, M. L. Lin, K. Matsuda, J. H. Park, T. Ueki, T. Ohta, K. Hirata, M. Fukuda, Y. Nakamura, and T. Katagiri. 2008. "Involvement of kinesin family member 2C/mitotic centromere-associated kinesin overexpression in mammary carcinogenesis." *Cancer Sci* 99 (1):62–70. doi:10.1111/j.1349-7006.2007.00635.x.

Shipley, K., M. Hekmat-Nejad, J. Turner, C. Moores, R. Anderson, R. Milligan, R. Sakowicz, and R. Fletterick. 2004. "Structure of a kinesin microtubule depolymerization machine." *EMBO J* 23 (7):1422–32. doi:10.1038/sj.emboj.7600165

Srayko, M., A. Kaya, J. Stamford, and A. A. Hyman. 2005. "Identification and characterization of factors required for microtubule growth and nucleation in the early C. elegans embryo." *Dev Cell* 9 (2):223–36. doi:10.1016/j.devcel.2005.07.003.

Talapatra, S. K., B. Harker, and J. P. Welburn. 2015. "The C-terminal region of the motor protein MCAK controls its structure and activity through a conformational switch." *Elife* 4. doi:10.7554/eLife.06421.

Tan, D., A. B. Asenjo, V. Mennella, D. J. Sharp, and H. Sosa. 2006. "Kinesin-13s form rings around microtubules." *J Cell Biol* 175 (1):25–31. doi:jcb.200605194 [pii] 10.1083/jcb.200605194.

Tanenbaum, M. E., L. Macurek, B. van der Vaart, M. Galli, A. Akhmanova, and R. H. Medema. 2011a. "A complex of Kif18b and MCAK promotes microtubule depolymerization and is negatively regulated by Aurora kinases." *Curr Biol* 21 (16):1356–65. doi:10.1016/j.cub.2011.07.017.

Tanenbaum, M. E., R. H. Medema, and A. Akhmanova. 2011b. "Regulation of localization and activity of the microtubule depolymerase MCAK." *BioArchitecture* 1:80–7.

Tanno, Y., T. S. Kitajima, T. Honda, Y. Ando, K. Ishiguro, and Y. Watanabe. 2010. "Phosphorylation of mammalian Sgo2 by Aurora B recruits PP2A and MCAK to centromeres." *Genes Dev* 24 (19):2169–79. doi:10.1101/gad.1945310.

Tian, G., A. G. Cristancho, H. A. Dubbs, G. T. Liu, N. J. Cowan, and E. M. Goldberg. 2016. "A patient with lissencephaly, developmental delay, and infantile spasms, due to de novo heterozygous mutation of KIF2A." *Mol Genet Genomic Med* 4 (6):599–603. doi:10.1002/mgg3.236.

Trofimova, D., M. Paydar, A. Zara, L. Talje, B. H. Kwok, and J. S. Allingham. 2018. "Ternary complex of Kif2A-bound tandem tubulin heterodimers represents a kinesin-13-mediated microtubule depolymerization reaction intermediate." *Nat Commun* 9 (1):2628. doi:10.1038/s41467-018-05025-7.

Uehara, R., Y. Tsukada, T. Kamasaki, I. Poser, K. Yoda, D. W. Gerlich, and G. Goshima. 2013. "Aurora B and Kif2A control microtubule length for assembly of a functional central spindle during anaphase." *J Cell Biol* 202 (4):623–36. doi:10.1083/jcb.201302123.

van Heesbeen, Rghp, J. A. Raaijmakers, M. E. Tanenbaum, V. A. Halim, D. Lelieveld, C. Lieftink, A. J. R. Heck, D. A. Egan, and R. H. Medema. 2017. "Aurora A, MCAK, and Kif18b promote Eg5-independent spindle formation." *Chromosoma* 126 (4):473–86. doi:10.1007/s00412-016-0607-4.

Walczak, C. E., E. C. Gan, A. Desai, T. J. Mitchison, and S. L. Kline-Smith. 2002. "The microtubule-destabilizing kinesin XKCM1 is required for chromosome positioning during spindle assembly." *Curr Biol* 12 (21):1885–9. doi:S0960982202012277 [pii].

Walczak, C. E., T. J. Mitchison, and A. Desai. 1996. "XKCM1: a Xenopus kinesin-related protein that regulates microtubule dynamics during mitotic spindle assembly." *Cell* 84 (1):37–47.

Wang, D., H. Zhu, Q. Ye, C. Wang, and Y. Xu. 2016. "Prognostic value of KIF2A and HER2-Neu overexpression in patients with epithelial ovarian cancer." *Medicine (Baltimore)* 95 (8):e2803. doi:10.1097/MD.0000000000002803.

Wang, W., S. Cantos-Fernandes, Y. Lv, H. Kuerban, S. Ahmad, C. Wang, and B. Gigant. 2017. "Insight into microtubule disassembly by kinesin-13s from the structure of Kif2C bound to tubulin." *Nat Commun* 8 (1):70. doi:10.1038/s41467-017-00091-9.

Welburn, J. P., and I. M. Cheeseman. 2012. "The microtubule-binding protein Cep170 promotes the targeting of the kinesin-13 depolymerase Kif2b to the mitotic spindle." *Mol Biol Cell* 23 (24):4786–95. doi:10.1091/mbc.E12-03-0214.

Wickstead, B., K. Gull, and T. A. Richards. 2010. "Patterns of kinesin evolution reveal a complex ancestral eukaryote with a multifunctional cytoskeleton." *BMC Evol Biol* 10:110. doi:10.1186/1471-2148-10-110.

Wilbur, J. D., and R. Heald. 2013. "Mitotic spindle scaling during Xenopus development by kif2a and importin alpha." *Elife* 2:e00290. doi:10.7554/eLife.00290.

Wordeman, L., and T. J. Mitchison. 1995. "Identification and partial characterization of mitotic centromere-associated kinesin, a kinesin-related protein that associates with centromeres during mitosis." *J Cell Biol* 128 (1–2):95–104.

Wordeman, L., M. Wagenbach, and G. von Dassow. 2007. "MCAK facilitates chromosome movement by promoting kinetochore microtubule turnover." *J Cell Biol* 179 (5):869–79. doi:jcb.200707120 [pii] 10.1083/jcb.200707120.

Xie, T., X. Li, F. Ye, C. Lu, H. Huang, F. Wang, X. Cao, and C. Zhong. 2018. "High KIF2A expression promotes proliferation, migration and predicts poor prognosis in lung adenocarcinoma." *Biochem Biophys Res Commun* 497 (1):65–72. doi:10.1016/j.bbrc.2018.02.020.

Yi, Z. Y., X. S. Ma, Q. X. Liang, T. Zhang, Z. Y. Xu, T. G. Meng, Y. C. Ouyang, Y. Hou, H. Schatten, Q. Y. Sun, and S. Quan. 2016. "Kif2a regulates spindle organization and cell cycle progression in meiotic oocytes." *Sci Rep* 6:38574. doi:10.1038/srep38574.

Zhang, L., H. Shao, Y. Huang, F. Yan, Y. Chu, H. Hou, M. Zhu, C. Fu, F. Aikhionbare, G. Fang, X. Ding, and X. Yao. 2011. "PLK1 phosphorylates mitotic centromere-associated kinesin and promotes its depolymerase activity." *J Biol Chem* 286 (4):3033–46. doi:10.1074/jbc.M110.165340.

Zhang, X., S. C. Ems-McClung, and C. E. Walczak. 2008. "Aurora A phosphorylates MCAK to control ran-dependent spindle bipolarity." *Mol Biol Cell* 19 (7):2752–65. doi:10.1091/mbc.E08-02-0198.

Zhang, X., C. Ma, Q. Wang, J. Liu, M. Tian, Y. Yuan, X. Li, and X. Qu. 2016. "Role of KIF2A in the progression and metastasis of human glioma." *Mol Med Rep* 13 (2):1781–7. doi:10.3892/mmr.2015.4700.

Zhang, Y., X. You, H. Liu, M. Xu, Q. Dang, L. Yang, J. Huang, and W. Shi. 2017. "High KIF2A expression predicts unfavorable prognosis in diffuse large B cell lymphoma." *Ann Hematol* 96 (9):1485–91. doi:10.1007/s00277-017-3047-1.

Zong, H., S. K. Carnes, C. Moe, C. E. Walczak, and S. C. Ems-McClung. 2016. "The far C-terminus of MCAK regulates its conformation and spindle pole focusing." *Mol Biol Cell* 27 (9):1451–64. doi:10.1091/mbc.E15-10-0699.

10 The Kinesin-14 Family

Marcus Braun, Stefan Diez and Zdenek Lansky

CONTENTS

The Kinesin-14s are microtubule crosslinkers and tip-trackers, unique among kinesins in their microtubule minus-end-directed motility.

10.1 EXAMPLE FAMILY MEMBERS

Mammalian: HSET (KIFC1)
Drosophila melanogaster: Ncd
Xenopus laevis: XCTK2 (KIFC1)
Schizosaccharomyces pombe: KLP2, Pkl1
Saccharomyces cerevisiae: Kar3

10.2 STRUCTURAL INFORMATION

10.2.1 HOMODIMERIC KINESIN-14S

Kinesin-14 family members are homodimeric molecules, with the noteworthy exception of the heterodimers found in budding yeast. The motor domain is located

115

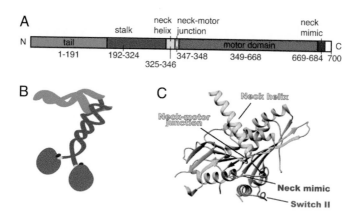

FIGURE 10.1 Structure of Kinesin-14 Ncd. (A) Protein domains, (B) dimeric molecule, and (C) 3D structure of Ncd catalytic core (PDB: 3U06, chain B). Copyright Yamagishi et al. (2016).

at the C-terminal end of the primary sequence, preceded by a coiled-coil region (stalk domain) through which dimerisation occurs (Figure 10.1A, B). They typically possess an additional motor domain-independent microtubule-binding site in their N-terminal tail-domain (Chandra et al. 1993), which interacts electrostatically with microtubules through a patch of positively charged amino acids (Furuta and Toyoshima 2008), allowing for microtubule–microtubule crosslinking and sliding (Fink et al. 2009; Braun et al. 2009).

Distinctively, the Kinesin-14 motor domain is located at the C-terminus of the sequence and, despite having reversed directionality, the Kinesin-14 motor domain is homologous in sequence and structure to the canonical Kinesin-1 motor domain (Sablin et al. 1998) and binds to the same microtubule-binding site, in the same orientation (Hirose et al. 1996; Hoenger et al. 1995). Consistent with this finding, directionality is not determined by the motor domain (Case et al. 1997); instead, directionality is controlled by a 'neck region' located between the stalk and the motor domain (Endow and Waligora 1998; Sablin et al. 1998), which consists of a 'neck helix' and a 'neck-motor junction' on the N-terminal side of the motor domain. On the C-terminal side of the motor domain, the position of the 'neck linker' in many other kinesin families, including Kinesin-1, Kinesin-14s possess a conserved, five-amino acid long 'neck-mimic', which displays similarity to the Kinesin-1 neck linker. By directly interacting with the neck helix, the neck-mimic is required for generating minus-end-directed motility (Yamagishi et al. 2016) (Figure 10.1 A–C). Cryogenic electron microscopy (cryo-EM) of Ncd revealed the nucleotide-dependent conformational changes of the neck helix (Wendt et al. 2002; Endres et al. 2005), involving the minus-end-directed swing of a lever-arm mechanical element (Wendt et al. 2002; Yun et al. 2003; Endres et al. 2005; Hirose et al. 2006; Makino et al. 2007; Liu, Pemble, and Endow 2012). Evidence also exists for an electrostatically guided diffusion-to-capture of the Ncd motor domain by microtubules, which is sufficient to result in directionally biased binding (Grant et al. 2011).

10.2.2 HETERODIMERIC KINESIN-14S

A noteworthy and unique exception to the typically homodimeric Kinesin-14 family members is the presence of Kinesin-14 heterodimers in budding yeast. Kar3 dimerises with Vik1, which, despite not being conserved at the level of the DNA sequence, adopts the fold typical of a kinesin motor domain, although it lacks a nucleotide-binding site (Allingham et al. 2007). Nevertheless, the heterodimer is a fully functional molecular motor (Yamagishi et al. 2016; Rank et al. 2012; Duan et al. 2012) and structural changes, similar to those described above for homodimeric Kinesin-14s, are observed for the non-catalytic Vik1 subunit (Duan et al. 2012; Gonzalez et al. 2013).

10.3 FUNCTIONAL PROPERTIES

10.3.1 DIRECTIONALITY

Members of the Kinesin-14 family are typically translocating motors that support microtubule motility. The best-described member of the family is the *Drosophila* Ncd (non-claret disjunctional), which owes its name to a *Drosophila* mutant defective in chromosome segregation that was discovered almost a hundred years ago (Sturtevant 1929). The locus was later found to encode a kinesin-like protein (Endow, Henikoff, and Soler-Niedziela 1990; McDonald and Goldstein 1990), which, unlike any other kinesin, moved towards the minus ends of microtubules (McDonald, Stewart, and Goldstein 1990; Walker, Salmon, and Endow 1990), and which was eventually classified in 2004 as Kinesin-14 (Lawrence et al. 2004). Ncd, like all Kinesin-14s, possesses the ubiquitously conserved kinesin motor domain. Specific interaction between the Ncd neck and motor domain are crucial for movement towards the microtubule minus end (Endow and Waligora 1998; Sablin et al. 1998), and a single amino acid change in the Ncd neck causes the motor to become bidirectional (Endow and Higuchi 2000). That small changes are sufficient to alter the directionality of Ncd is consistent with optical trapping experiments showing that 30% of power strokes generated by wild-type Ncd are directed towards the plus end of the microtubule (Butterfield et al. 2010).

10.3.2 PROCESSIVITY

As single molecules, Kinesin-14 motors, again exemplified by *Drosophila* Ncd, are mostly non-processive (Figure 10.2A) (Case et al. 1997; Foster and Gilbert 2000), and transport becomes processive only when multiple motors cooperate. The latter behaviour has been demonstrated in gliding microtubule motility assays, where the minimal distance of smoothly gliding microtubule movement was found to be dependent on motor density (Stewart, Semerjian, and Schmidt 1998), and also in experiments, where the run length of DNA scaffolds was found to be dependent on the number of motors coupled to them (Furuta et al. 2013). Coupling of multiple Kinesin-14 motors was also shown to enable processive transport of intracellular cargo along microtubules in plants (Jonsson et al. 2015) and of Pkl1 functionalised quantum dots (Furuta et al. 2008). Clustering of the Kinesin-14, HSET/KIFC1, via binding of unpolymerised

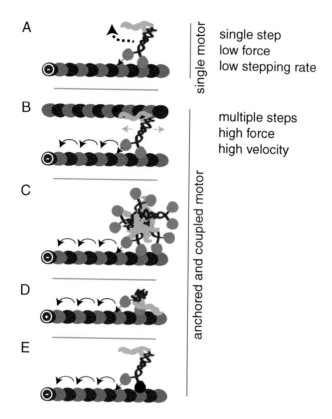

FIGURE 10.2 Enabling Kinesin-14 processivity by anchoring to microtubules. (A) un-processive, non-anchored Kinesin-14, (B) Kinesin-14 anchored via its tail to a crosslinked microtubule in an overlap (Fink et al. 2009), (C) multiple Kinesin-14s in a cluster, anchoring each other (Norris et al. 2018), (D) KLPA anchored via its tail to one microtubule (Popchock et al. 2017), and (E) Kar3 anchored via Vik1 (black) (Allingham et al. 2007).

tubulin, also increases processivity (Figure 10.2C) (Norris et al. 2018). Whilst, some Kinesin-14s, such as the plant kinesin OsKCH2, have been reported to be intrinsically processive, minus-end-directed motors (Tseng et al. 2018). Interestingly, kinesin-like protein A (KLPA) (a Kinesin-14 from *Aspergillus nidulans*) has recently been reported to be able to switch from the canonical non-processive, minus-end-directionality to processive, plus-end-directionality when its tail is bound to the same microtubule as the motor domain (Figure 10.2D) (Popchock et al. 2017).

A unique strategy for enabling kinesin processivity is seen in the budding yeast Kar3, which associates with either Cik1 or Vik1 to form heterodimers (Figure 10.2E) (Manning et al. 1999; Page et al. 1994). These accessory subunits bind to microtubules, but do not hydrolyse ATP (Allingham et al. 2007). They do, however, have a key impact on motility of the heterodimer. By providing a 'foothold' for the catalytic Kar3, they enable its processive movement towards the microtubule minus end (Mieck et al. 2015) – the identity of the partner in Kar3 heterodimers dictates their subcellular localisation and function (Manning et al. 1999; Gardner et al. 2008).

10.3.3 VELOCITY

As Kinesin-14s are typically non-processive, it is hard to define the velocity of an individual motor molecule. Rather, it makes more sense to characterise the velocity of cargo transported by multiple motors. Such cargo transport (or, equivalently, the gliding of microtubules on motor-coated surfaces) occurs at comparatively slow velocities, between 70 nm/s and 250 nm/s (McDonald, Stewart, and Goldstein 1990; Walker, Salmon, and Endow 1990; Furuta et al. 2013; Fink et al. 2009; Lüdecke et al. 2018). More recent experiments show microtubule gliding velocities in the ranges of 20–40nm/s and 30–70nm/s, depending on the motor density for XCTK2 (Hentrich and Surrey 2010) and HSET/KIFC1 (Braun et al. 2017), respectively. Interestingly, when connecting two to four Ncd motors onto DNA scaffolds at fixed distances from each other, no significant changes in transport velocity (~150 nm/s) were observed (Furuta et al. 2013). Thus, the transport velocity by Kinesin-14 motors appears to be dependent on motor density, but not motor number, likely due to steric effects or mechanical coupling. From measurements of the bulk ATPase activity, it can be inferred that the rate at which Kinesin-14 performs individual power strokes ranges from one to four per motor per second (Pechatnikova and Taylor 1999; Furuta and Toyoshima 2008; Braun et al. 2009; Szczesna and Kasprzak 2012). This is in line with single-molecule optical trapping evidence, showing that an approximately 9-nm displacement of the motor domain occurs over a period of 200–400 ms, independent of ATP concentration (deCastro et al. 2000). Taken together, these estimates would translate to a 'single-motor velocity' of about 10–40 nm/s. So far, it is not clear by which mechanism, for example, just two coupled motors can reach velocities of up to 150 nm/s (Furuta et al. 2013).

10.3.4 CHEMO-MECHANICAL CYCLE

Although Kinesin-14s are dimeric motor proteins, it is believed that, in each cycle, only one of the two motor domains typically mediates force generation (Hirose et al. 1996; Endres et al. 2005). In solution, both motor domains are bound to ADP and interact only weakly with the microtubule, as evidenced by biochemical (Crevel, Lockhart, and Cross 1996; Pechatnikova and Taylor 1997; Rosenfeld et al. 1996), electron microscopy (Hirose, Cross, and Amos 1998; Wendt et al. 2002) and single-molecule optical microscopy studies (Furuta et al. 2008). Upon ADP release from one of the motor domains, this motor binds strongly to the microtubule in the absence of nucleotide (Hirose, Cross, and Amos 1998; Wendt et al. 2002; Endres et al. 2005). After binding ATP, it remains firmly attached to the microtubule (Endres et al. 2005) and detaches only after ATP hydrolysis (Foster, Correia, and Gilbert 1998).

For the working stroke of a single Ncd motor head, a lever arm model has been suggested (Endres et al. 2005; Yun et al. 2003). It is currently assumed that the working stroke of Ncd comprises two substeps: an initial small movement of its stalk in a lateral direction (i.e. off-axis to the orientation of the microtubule) when ADP is released, and a second, main component in a longitudinal direction (i.e. parallel to the orientation of the microtubule) upon ATP binding (Hallen, Liang, and Endow 2011; Nitzsche et al. 2016). The off-axis component in the power stroke is believed to be responsible for the Ncd's capability to not only translocate microtubules forward but also to rotate them around their own axes in gliding motility assays, with right-handed pitches between 300 nm and 2 μm in an ATP-dependent manner (Walker, Salmon, and Endow 1990; Nitzsche et al. 2016).

10.3.5 Force Generation

Low forces were first applied to single Ncd motors in a three-bead, optical-trap assay in order to resolve the motor's mechanochemical cycle, comprising the above mentioned 9-nm displacement of the motor domain (deCastro et al. 2000). Further trapping experiments with Ncd motors coupled to DNA scaffolds determined that the average maximum force of multiple Ncds increased additively with motor number (Furuta et al. 2013). Though the force generated by a single Ncd motor was too weak to be measured reliably, extrapolation of the data from two, three and four motors yielded single-motor forces of about 0.4 pN. This force range was confirmed in gliding motility assays, where an optical trap was used to determine the maximum forces of bead-loaded microtubules gliding on surfaces coated with Ncd motors (Lüdecke et al. 2018).

10.3.6 Microtubule–microtubule Crosslinking

A number of Kinesin-14s (such as Ncd, KLP14, HSET/KIFC1 and XCTK2) exhibit a second microtubule-binding site in the tail domain. This additional binding site has a nucleotide-independent diffusive interaction with microtubules (Figure 10.3A) (Furuta and Toyoshima 2008; Fink et al. 2009). It is the existence of this second microtubule-binding site which allows these kinesins to crosslink and slide two microtubules (Oladipo, Cowan, and Rodionov 2007). When bound between two microtubules *in vitro*, Kinesin-14 slides antiparallel microtubules while stationarily crosslinking parallel microtubules (Figure 10.2B, C) (Braun et al. 2009; 2017; Fink et al. 2009; Hentrich and Surrey 2010; Oladipo, Cowan, and Rodionov 2007). However, the sliding velocity of crosslinked antiparallel microtubules is significantly lower than the gliding velocity of microtubules propelled by motors anchored to a rigid surface, such as a glass cover slip. This finding is attributed to (i) the steric hindrance between motors present at high densities between two microtubules (Braun et al. 2017) and (ii) the diffusive anchorage of the motor tails to the second microtubule, causing motor slippage (Lüdecke et al. 2018), similar to motors anchored diffusively in a lipid bilayer (Grover et al. 2016). The same arguments hold true for a significant reduction in the generated force as estimated from force-balance experiments involving Kinesin-5 acting antagonistically to Kinesin-14 (Civelekoglu-Scholey et al. 2010; Hentrich and Surrey 2010; Roostalu et al. 2018; Tao et al. 2006) and directly measured in an optical trap (Lüdecke et al. 2018). Beyond motorized microtubule–microtubule sliding, Kinesin-14 motors can also give rise to entropic forces in microtubule overlaps to prevent overlap separation (Braun et al. 2017), following a mechanism similar to that recently reported for the passive, diffusive crosslinker Ase1 confined between two microtubules (Lansky et al. 2015). In microtubule networks, Kinesin-14 motors alone are thus expected to fulfil a multitude of functionalities: (i) active sliding due to ensemble force generation, (ii) adaptive

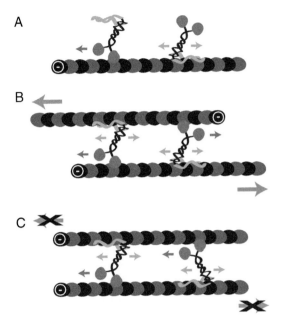

FIGURE 10.3 Modes of Kinesin-14 interaction with microtubules. (A) non-processive, single-step, walking (left) and unbiased diffusion (right), (B) sliding of antiparallel overlapping microtubules, and (C) locking and stabilising parallel overlapping microtubules.

force regulation via tail slippage, motor-density effects and force balance with other motors, as well as (iii) entropic force generation to avoid the separation of microtubule–microtubule overlaps.

Interestingly, the rice Kinesin-14 OsKCH1 does not bind to a second microtubule but rather to actin filaments, transporting those along microtubules with two distinct velocities, depending on their orientation relative to the microtubule (Walter et al. 2015). In a way, the Kinesin-14-mediated, continuous sliding of filaments (microtubules and actin filaments both serving as 'cargo') along each other can be seen as motor coupling to achieve processive motility.

10.4 PHYSIOLOGICAL ROLE

10.4.1 Roles in Mitosis and Meiosis

Kinesin-14 localises to mitotic and meiotic spindles, where it is enriched at the spindle poles (Hatsumi and Endow, 1992b; Walczak, Verma, and Mitchison, 1997) and is involved in cell division, primarily in the focusing of the poles (Figure 10.4, upper left panel), in maintenance of mitotic and meiotic spindle shape, length and bipolarity, positioning of centrosomes and correct chromosome distribution. Within the spindle, Kinesin-14 is spatially regulated by the gradient of Ran-GTP by modulating the affinity of its microtubule-binding tail domain for microtubules (Weaver et al. 2015; Ems-McClung, Zheng, and Walczak 2004).

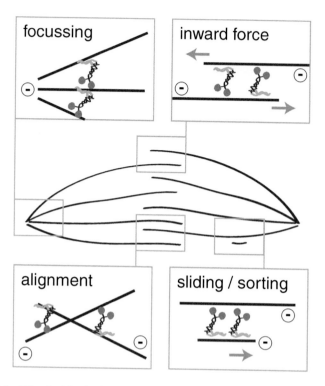

FIGURE 10.4 Kinesin-14 roles in mitosis are (upper left) pole focusing, (upper right) generation of spindle-inward forces, (lower left) microtubule alignment and (lower right) parallel microtubule sliding.

Kinesin-14 is especially important for the formation of spindles in the absence of centrosomes (Mountain et al. 1999). *In vitro*, in an extract of *X. laevis* eggs, which lack centrosomes, Kinesin-14 XCTK2 focuses the microtubule minus ends to mediate mitotic spindle assembly (Matthies et al. 1996; Walczak et al. 1998). Similar observations were obtained in other acentrosomal systems, such as in *Drosophila* oocytes in meiosis I and II (Hatsumi and Endow 1992; Endow and Komma 1997) and in mitotic spindles in plants (Ambrose and Cyr 2007). In early stages of mitosis, Kinesin-14 also mediates the focusing of microtubules nucleating from chromosomes and, dependent upon EB1, their capture by and alignment with the microtubules nucleated from centrosomes (Goshima, Nédélec, and Vale 2005). The ability of Kinesin-14 to focus minus ends of microtubules into poles, the microtubules forming asters, has been indeed reconstituted in a minimal system consisting of only Kinesin-14 motors and microtubules, showing that Kinesin-14 is an autonomous pole-forming motor (Hentrich and Surrey 2010).

In terms of spindle pole focusing, Kinesin-14 has been reported to interact with various pole-associated factors. Its interaction with γ-TuRC has been reported in several organisms. The *S. pombe* Kinesin-14, Pkl1, interacts with γ-TuRC directly (Rodriguez et al. 2008) and regulates microtubule nucleation (Olmsted et al. 2014).

To focus the spindle poles, Pkl1 also cooperates with the mitotic spindle disanchored protein 1 (Msd1) and another spindle anchoring factor Wdr8 (Yukawa, Ikebe, and Toda 2015). The absence of Pkl1p results in the defocusing of the poles and a defect in a checkpoint that monitors chromosome biorientation, resulting in frequent precocious anaphase and aneuploidy (Grishchuk, Spiridonov, and McIntosh 2007; Syrovatkina and Tran 2015). In *D. melanogaster* oocytes, γ-tubulin and Ncd enhance the stable association of the augmin complex with the spindle poles to facilitate microtubule organisation and chromosome congression (Colombié et al. 2013). The human Kinesin-14, HSET, is involved (together with Kinesin-5 and dynein) in poleward transport of non-centrosomal mitotic microtubules through interaction with γ-TuRC, which results in the accumulation of their minus ends at the pole (Lecland and Lüders 2014). Inhibition of HSET leads to the disruption of the microtubule aster assembly. However, simultaneous inhibition of the Kinesin-5, Eg5, restores the microtubule aster formation, similarly to centrosome separation and spindle organisation, suggesting that Kinesin-14 and Kinesin-5 are playing opposing roles in these processes (Mountain et al. 1999).

Indeed, experiments in various organisms show that Kinesin-14, together with other microtubule-sliding motors, maintain the length of the mitotic spindle. Experiments in yeast reveal that the action of the antagonistic motors Kinesin-14 and Kinesin-5 determine the spindle length (Saunders, Lengyel, and Hoyt 1997). Control of the spindle length is mediated by the plus- and minus-end-directed mechanical forces exerted by these motors, which crosslink and slide microtubules. By sliding antiparallel microtubules, Kinesin-14 generates inward forces on the spindle poles (Figure 10.4, upper right panel), which is opposed by Kinesin-5, generating outward force (Sharp et al. 1999; 2000). Experiments like these led to the generation of a force-balance model of prometaphase spindle formation, whereby Kinesin-14 and Kinesin-5 generate opposing power strokes with load-dependent detachment. The forces exerted by the ensembles of these motors thus maintain the correct distance between the poles (Civelekoglu-Scholey et al. 2010). The force-balance model is supported by experiments showing that changing the ratio between the two motors affects the spindle length (Brust-Mascher et al. 2009) and that simultaneous inhibition of Kinesin-14 and Kinesin-5 restores the separation of centrosomes, showing also that additional motors and spindle mechanisms cooperate to form spindles when the Kinesin-14–Kinesin-5 pair is removed (Mountain et al. 1999). Intriguingly, the deletion of Kinesin-14 in various organisms often results in shorter spindle lengths (Cai et al. 2009; Hepperla et al. 2014), contrasting with the notion that Kinesin-14 is an inward force generator. This suggests that Kinesin-14 primarily aligns antiparallel microtubules along the spindle axis during metaphase, thus allowing Kinesin-5 to then exert outward force (Figure 10.4, lower left panel) (Gardner et al. 2008). Alternatively, Kinesin-14 might slide parallel microtubules, moving the poles outward (Figure 10.4, lower right panel) (Cai et al. 2009). Mechanistically, such parallel sliding could occur when the two microtubules in the crosslinked pair are not identical, with respect to the motor binding, which breaks the symmetry of the sliding process (Fink et al. 2009). We can only speculate what would cause this difference, e.g. tubulin post-translational modifications or other factors, such as a component of the augmin or γ-TuRC complex.

10.4.2 ROLES IN PLANTS

In most eukaryotic organisms, microtubule minus-end-directed motility is driven by dynein and Kinesin-14. However, in plants, dynein is not present and Kinesin-14 family members are thus the only minus-end-directed motors. This may be the reason for the multiple Kinesin-14 genes in plants (Reddy and Day 2001). As in animals and fungi, Kinesin-14s in plants play a role in maintaining the spindle morphology (Ambrose and Cyr 2007). Interestingly, no plant Kinesin-14 has been found to be solely required for proper chromosome segregation. This suggests that either several Kinesin-14s are functionally redundant or that there are other redundant pathways leading to proper spindle formation in plants (Gicking et al. 2018). Due to the lack of dynein in plants, Kinesin-14s have various other roles, mostly in transport. They are involved in nuclear migration (Frey, Klotz, and Nick 2010; Yamada and Goshima 2018; Yamada et al. 2017) and minus-end-directed transport of cargoes, such as mitochondria (Yang et al. 2011; Yamada et al. 2017) and chloroplasts (Suetsugu et al. 2010), and they move neocentromeres along spindle microtubules to cause meiotic drive in maize (Dawe et al. 2018).

10.4.3 OTHER ROLES

Apart from their roles in cell division and their prominent role as the only minus-end transporter in plants, Kinesin-14s have been reported to be important in several other cellular systems. In *S. pombe*, the Kinesin-14, KLP2, is involved in the formation and maintenance of the interphase microtubule array (Carazo-Salas and Nurse 2006; Daga et al. 2006), through localising to plus tips of newly nucleated microtubules and sliding them along the pre-existing ones to the correct positions (Janson et al. 2007). In *S. cerevisiae*, the Kinesin-14, Kar3, drives lateral sliding of kinetochores along the side of microtubules (Tanaka et al. 2007; 2005). Moreover, Kinesin-14 is involved in structural integrity and positioning of the Golgi complex in non-polarised mammalian cells (She et al. 2017) and in DNA repair (Chung et al. 2015).

10.5 INVOLVEMENT IN DISEASE

Kinesin-14 is not essential for mitosis in somatic cells, and mutants of *Drosophila* Kinesin-14 Ncd develop normally with few mitotic defects (Basto et al. 2008). However, in cells with multiple centrosomes, as often found in cancer cells (Nigg, Čajánek, and Arquint 2014; Anderhub, Krämer, and Maier 2012; Ganem, Godinho, and Pellman 2009), Ncd increases the efficiency of centrosome clustering, demonstrating that, in this context, Kinesin-14 is important for the formation of bipolar spindles (Basto et al. 2008). Cells with multiple centrosomes can form multipolar spindles during mitosis, which can cause aneuploidy, resulting in cell death, unless the centrosomes cluster together to form two prominent poles ('pseudopoles'), which then assemble bipolar mitotic spindles allowing the cells to progress through bipolar mitosis, albeit with increased frequency of lagging chromosomes (Ring 1982;

Ganem, Godinho, and Pellman 2009; Basto et al. 2008). Indeed, knockdown of Kinesin-14 induces mitotic defects in cancer cells with multiple centrosomes, presumably due to reduced pseudopole formation and focusing, resulting in increased cell death through multipolar anaphase (Kwon et al. 2008). HSET has also been found to be required for the survival of cancer cells that have the regular two centrosomes, suggesting that it plays a more complex role in cancer cells (Kleylein-Sohn et al. 2013). Indeed, HSET expression is up-regulated in cancer cells and its high expression levels correlate with metastasis (Wu et al. 2013; Li et al. 2015; Grinberg-Rashi et al. 2009). HSET is thus an attractive target for anti-cancer drugs as its inhibition, in contrast to other mitotic kinesins, would result in the inhibition of centrosome clustering, selectively targeting only multi-centrosomal cancer cells. Several specific HSET inhibitors preventing centrosome clustering have been synthesised to date and their differential effect on cancer cells with multiple centrosomes as compared to normal cells has been demonstrated (Watts et al. 2013; Wu et al. 2013; Yang et al. 2014; Zhang et al. 2016). These inhibitors provide both a tool for the investigation of the role of HSET in mitosis and cancer and a starting point for the development of future therapeutics.

REFERENCES

Allingham, John S, Lisa R Sproul, Ivan Rayment, and Susan P Gilbert. 2007. "Vik1 Modulates Microtubule-Kar3 Interactions Through a Motor Domain That Lacks an Active Site." *Cell* 128 (6): 1161–72. doi:10.1016/j.cell.2006.12.046.

Ambrose, J Christian, and Richard Cyr. 2007. "The Kinesin ATK5 Functions in Early Spindle Assembly in Arabidopsis." *The Plant Cell* 19 (1): 226–36. doi:10.1105/tpc.106.047613.

Anderhub, Simon J, Alwin Krämer, and Bettina Maier. 2012. "Centrosome Amplification in Tumorigenesis." *Cancer Letters* 322 (1): 8–17. doi:10.1016/j.canlet.2012.02.006.

Basto, Renata, Kathrin Brunk, Tatiana Vinadogrova, Nina Peel, Anna Franz, Alexey Khodjakov, and Jordan W Raff. 2008. "Centrosome Amplification Can Initiate Tumorigenesis in Flies." *Cell* 133 (6): 1032–42. doi:10.1016/j.cell.2008.05.039.

Braun, Marcus, Douglas R Drummond, Robert A Cross, and Andrew D McAinsh. 2009. "The Kinesin-14 Klp2 Organizes Microtubules Into Parallel Bundles by an ATP-Dependent Sorting Mechanism." *Nature Cell Biology* 11 (6): 724–30. doi:10.1038/ncb1878.

Braun, Marcus, Zdenek Lansky, Agata Szuba, Friedrich W Schwarz, Aniruddha Mitra, Mengfei Gao, Annemarie Lüdecke, Pieter Rein ten Wolde, and Stefan Diez. 2017. "Changes in Microtubule Overlap Length Regulate Kinesin-14-Driven Microtubule Sliding." *Nature Chemical Biology* 13 (12): 1245–52. doi:10.1038/nchembio.2495.

Brust-Mascher, Ingrid, Patrizia Sommi, Dhanya K Cheerambathur, and Jonathan M Scholey. 2009. "Kinesin-5–Dependent Poleward Flux and Spindle Length Control in Drosophila Embryo Mitosis." Edited by Yixian Zheng. *Molecular Biology of the Cell* 20 (6): 1749–62. doi:10.1091/mbc.e08-10-1033.

Butterfield, Anthony E, Russell J Stewart, Christoph F Schmidt, and Mikhail Skliar. 2010. "Bidirectional Power Stroke by Ncd Kinesin." *Biophysical Journal* 99 (12): 3905–15. doi:10.1016/j.bpj.2010.10.045.

Cai, Shang, Lesley N Weaver, Stephanie C Ems-McClung, and Claire E Walczak. 2009. "Kinesin-14 Family Proteins HSET/XCTK2 Control Spindle Length by Cross-Linking and Sliding Microtubules." Edited by Kerry S Bloom. *Molecular Biology of the Cell* 20 (5): 1348–59. doi:10.1091/mbc.e08-09-0971.

Carazo-Salas, Rafael E, and Paul Nurse. 2006. "Self-Organization of Interphase Microtubule Arrays in Fission Yeast." *Nature Cell Biology* 8 (10): 1102–7. Nature Publishing Group. doi:10.1038/ncb1479.

Case, R B, D W Pierce, N Hom-Booher, C L Hart, and R D Vale. 1997. "The Directional Preference of Kinesin Motors Is Specified by an Element Outside of the Motor Catalytic Domain." *Cell* 90 (5): 959–66.

Chandra, R, E D Salmon, H P Erickson, A Lockhart, and S A Endow. 1993. "Structural and Functional Domains of the Drosophila Ncd Microtubule Motor Protein." *Journal of Biological Chemistry* 268 (12): 9005–13.

Chung, Daniel K C, Janet N Y Chan, Jonathan Strecker, Wei Zhang, Sasha Ebrahimi-Ardebili, Thomas Lu, Karan J Abraham, Daniel Durocher, and Karim Mekhail. 2015. "Perinuclear Tethers License Telomeric DSBs for a Broad Kinesin- and NPC-Dependent DNA Repair Process." *Nature Communications* 6 (1): 7742. Nature Publishing Group. doi:10.1038/ncomms8742.

Civelekoglu-Scholey, Gul, Li Tao, Ingrid Brust-Mascher, Roy Wollman, and Jonathan M Scholey. 2010. "Prometaphase Spindle Maintenance by an Antagonistic Motor-Dependent Force Balance Made Robust by a Disassembling Lamin-B Envelope." *The Journal of Cell Biology* 188 (1): 49–68. doi:10.1083/jcb.200908150.

Colombié, Nathalie, A Agata Głuszek, Ana M Meireles, and Hiroyuki Ohkura. 2013. "Meiosis-Specific Stable Binding of Augmin to Acentrosomal Spindle Poles Promotes Biased Microtubule Assembly in Oocytes." Edited by R Scott Hawley. *PLOS Genetics* 9 (6): e1003562–8. doi:10.1371/journal.pgen.1003562.

Crevel, Isabelle M T C, Andrew Lockhart, and Robert A Cross. 1996. "Weak and Strong States of Kinesin and Ncd." *Journal of Molecular Biology* 257 (1): 66–76. doi:10.1006/jmbi.1996.0147.

Daga, Rafael R, Kyeng-Gea Lee, Scott Bratman, Silvia Salas-Pino, and Fred Chang. 2006. "Self-Organization of Microtubule Bundles in Anucleate Fission Yeast Cells." *Nature Cell Biology* 8 (10): 1108–13. Nature Publishing Group. doi:10.1038/ncb1480.

Dawe, R Kelly, Elizabeth G Lowry, Jonathan I Gent, Michelle C Stitzer, Kyle W Swentowsky, David M Higgins, Jeffrey Ross-Ibarra, et al. 2018. "A Kinesin-14 Motor Activates Neocentromeres to Promote Meiotic Drive in Maize." *Cell* 173 (4): 839–50.e18. doi:10.1016/j.cell.2018.03.009.

deCastro, M J, R M Fondecave, L A Clarke, C F Schmidt, and R J Stewart. 2000. "Working Strokes by Single Molecules of the Kinesin-Related Microtubule Motor Ncd." *Nature Cell Biology* 2 (10): 724–29. Nature Publishing Group. doi:10.1038/35036357.

Duan, Da, Zhimeng Jia, Monika Joshi, Jacqueline Brunton, Michelle Chan, Doran Drew, Darlene Davis, and John S Allingham. 2012. "Neck Rotation and Neck Mimic Docking in the Noncatalytic Kar3-Associated Protein Vik1." *Journal of Biological Chemistry* 287 (48): 40292–301. American Society for Biochemistry and Molecular Biology. doi:10.1074/jbc.M112.416529.

Ems-McClung, Stephanie C, Yixian Zheng, and Claire E Walczak. 2004. "Importin Alpha/Beta and Ran-GTP Regulate XCTK2 Microtubule Binding Through a Bipartite Nuclear Localization Signal." *Molecular Biology of the Cell* 15 (1): 46–57. doi:10.1091/mbc.e03-07-0454.

Endow, S A, S Henikoff, and L Soler-Niedziela. 1990. "Mediation of Meiotic and Early Mitotic Chromosome Segregation in Drosophila by a Protein Related to Kinesin." *Nature* 345 (6270): 81–3. Nature Publishing Group. doi:10.1038/345081a0.

Endow, Sharyn A, and Donald J Komma. 1997. "Spindle Dynamics During Meiosis in DrosophilaOocytes." *The Journal of Cell Biology* 137 (6): 1321–36. doi:10.1083/jcb.137.6.1321.

Endow, S A, and K W Waligora. 1998. "Determinants of Kinesin Motor Polarity." *Science* 281 (5380): 1200–2. American Association for the Advancement of Science. doi:10.1126/science.281.5380.1200.

Endow, S A, and H Higuchi. 2000. "A Mutant of the Motor Protein Kinesin That Moves in Both Directions on Microtubules." *Nature* 406 (6798): 913–6. Nature Publishing Group. doi:10.1038/35022617.

Endres, Nicholas F, Craig Yoshioka, Ronald A Milligan, and Ronald D Vale. 2005. "A Lever-Arm Rotation Drives Motility of the Minus-End-Directed Kinesin Ncd." *Nature* 439 (7078): 875–8. doi:10.1038/nature04320.

Fink, Gero, Lukasz Hajdo, Krzysztof J Skowronek, Cordula Reuther, Andrzej A Kasprzak, and Stefan Diez. 2009. "The Mitotic Kinesin-14 Ncd Drives Directional Microtubule–Microtubule Sliding." *Nature Cell Biology* 11 (6): 717–23. doi:10.1038/ncb1877.

Foster, K A, and S P Gilbert. 2000. "Kinetic Studies of Dimeric Ncd: Evidence That Ncd Is Not Processive." *Biochemistry* 39 (7): 1784–91. doi:10.1021/bi991500b.

Foster, K A, J J Correia, and S P Gilbert. 1998. "Equilibrium Binding Studies of Non-Claret Disjunctional Protein (Ncd) Reveal Cooperative Interactions Between the Motor Domains." *Journal of Biological Chemistry* 273 (52): 35307–18. American Society for Biochemistry and Molecular Biology. doi:10.1074/jbc.273.52.35307.

Frey, Nicole, Jan Klotz, and Peter Nick. 2010. "A Kinesin with Calponin-Homology Domain Is Involved in Premitotic Nuclear Migration." *Journal of Experimental Botany* 61 (12): 3423–37. doi:10.1093/jxb/erq164.

Furuta, Ken'ya, Masaki Edamatsu, Yurina Maeda, and Yoko Y Toyoshima. 2008. "Diffusion and Directed Movement: In Vitro Motile Properties of Fission Yeast Kinesin-14 Pkl1." *Journal of Biological Chemistry* 283 (52): 36465–73. American Society for Biochemistry and Molecular Biology. doi:10.1074/jbc.M803730200.

Furuta, Ken'ya, and Yoko Yano Toyoshima. 2008. "Minus-End-Directed Motor Ncd Exhibits Processive Movement That Is Enhanced by Microtubule Bundling in Vitro." *Current Biology* 18 (2): 152–7. doi:10.1016/j.cub.2007.12.056.

Furuta, Ken'ya, Akane Furuta, Yoko Y Toyoshima, Misako Amino, Kazuhiro Oiwa, and Hiroaki Kojima. 2013. "Measuring Collective Transport by Defined Numbers of Processive and Nonprocessive Kinesin Motors." *Proceedings of the National Academy of Sciences of the United States of America* 110 (2): 501–6. doi:10.1073/pnas.1201390110.

Ganem, Neil J, Susana A Godinho, and David Pellman. 2009. "A Mechanism Linking Extra Centrosomes to Chromosomal Instability." *Nature* 460 (7252): 278–82. Nature Publishing Group. doi:10.1038/nature08136.

Gardner, Melissa K, Julian Haase, Karthikeyan Mythreye, Jeffrey N Molk, MaryBeth Anderson, Ajit P Joglekar, Eileen T O'Toole, et al. 2008. "The Microtubule-Based Motor Kar3 and Plus End–Binding Protein Bim1 Provide Structural Support for the Anaphase Spindle." *The Journal of Cell Biology* 180 (1): 91–100. doi:10.1083/jcb.200710164.

Gicking, Allison M, Kyle W Swentowsky, R Kelly Dawe, and Weihong Qiu. 2018. "Functional Diversification of the Kinesin-14 Family in Land Plants." Edited by Wilhelm Just. *FEBS Letters* 592 (12): 1918–28. doi:10.1002/1873-3468.13094.

Gonzalez, Miguel A, Julia Cope, Katherine C Rank, Chun Ju Chen, Peter Tittmann, Ivan Rayment, Susan P Gilbert, and Andreas Hoenger. 2013. "Common Mechanistic Themes for the Powerstroke of Kinesin-14 Motors." *Journal of Structural Biology* 184 (2): 335–44. doi:10.1016/j.jsb.2013.09.020.

Goshima, Gohta, François Nédélec, and Ronald D Vale. 2005. "Mechanisms for Focusing Mitotic Spindle Poles by Minus End–Directed Motor Proteins." *The Journal of Cell Biology* 171 (2): 229–40. doi:10.1083/jcb.200505107.

Grant, Barry J, Dana M Gheorghe, Wenjun Zheng, Maria Alonso, Gary Huber, Maciej Dlugosz, J Andrew McCammon, and Robert A Cross. 2011. "Electrostatically Biased Binding of Kinesin to Microtubules." Edited by Jon M Scholey. *PLoS Biology* 9 (11): e1001207–11. doi:10.1371/journal.pbio.1001207.

Grinberg-Rashi, Helena, Efrat Ofek, Marina Perelman, Jozef Skarda, Pnina Yaron, Marián Hajdúch, Jasmin Jacob-Hirsch, et al. 2009. "The Expression of Three Genes in Primary Non–Small Cell Lung Cancer Is Associated with Metastatic Spread to the Brain." *Clinical Cancer Research* 15 (5): 1755–61. doi:10.1158/1078-0432.CCR-08-2124.

Grishchuk, Ekaterina L, Ilia S Spiridonov, and J Richard McIntosh. 2007. "Mitotic Chromosome Biorientation in Fission Yeast Is Enhanced by Dynein and a Minus-End-Directed, Kinesin-Like Protein." Edited by Kerry Bloom. *Molecular Biology of the Cell* 18 (6): 2216–25. doi:10.1091/mbc.e06-11-0987.

Grover, Rahul, Janine Fischer, Friedrich W Schwarz, Wilhelm J Walter, Petra Schwille, and Stefan Diez. 2016. "Transport Efficiency of Membrane-Anchored Kinesin-1 Motors Depends on Motor Density and Diffusivity." *Proceedings of the National Academy of Sciences* 113 (46): E7185–93. doi:10.1073/pnas.1611398113.

Hallen, Mark A, Zhang-Yi Liang, and Sharyn A Endow. 2011. "Two-State Displacement by the Kinesin-14 Ncd Stalk." *Biophysical Chemistry* 154 (2–3): 56–65. doi:10.1016/j.bpc.2011.01.001.

Hatsumi, M, and S A Endow. 1992. "Mutants of the Microtubule Motor Protein, Nonclaret Disjunctional, Affect Spindle Structure and Chromosome Movement in Meiosis and Mitosis." *Journal of Cell Science* 101 (Pt 3) (March): 547–59.

Hentrich, Christian, and Thomas Surrey. 2010. "Microtubule Organization by the Antagonistic Mitotic Motors Kinesin-5 and Kinesin-14." *The Journal of Cell Biology* 189 (3): 465–80. doi:10.1083/jcb.200910125.

Hepperla, Austin J, Patrick T Willey, Courtney E Coombes, Breanna M Schuster, Maryam Gerami-Nejad, Mark McClellan, Soumya Mukherjee, et al. 2014. "Minus-End-Directed Kinesin-14 Motors Align Antiparallel Microtubules to Control Metaphase Spindle Length." *Developmental Cell* 31 (1): 61–72. doi:10.1016/j.devcel.2014.07.023.

Hirose, K, A Lockhart, R A Cross, and L A Amos. 1996. "Three-Dimensional Cryoelectron Microscopy of Dimeric Kinesin and Ncd Motor Domains on Microtubules." *Proceedings of the National Academy of Sciences* 93 (18): 9539–44. National Academy of Sciences. doi:10.1073/pnas.93.18.9539.

Hirose, K, R A Cross, and L A Amos. 1998. "Nucleotide-Dependent Structural Changes in Dimeric NCD Molecules Complexed to Microtubules." Edited by R. Huber. *Journal of Molecular Biology* 278 (2): 389–400. doi:10.1006/jmbi.1998.1709.

Hirose, Keiko, Erika Akimaru, Toshihiko Akiba, Sharyn A Endow, and Linda A Amos. 2006. "Large Conformational Changes in a Kinesin Motor Catalyzed by Interaction with Microtubules." *Molecular Cell* 23 (6): 913–23. doi:10.1016/j.molcel.2006.07.020.

Hoenger, A, E P Sablin, R D Vale, R J Fletterick, and R A Milligan. 1995. "Three-Dimensional Structure of a Tubulin-Motor-Protein Complex." *Nature* 376 (6537): 271–4. Nature Publishing Group. doi:10.1038/376271a0.

Janson, Marcel E, Rose Loughlin, Isabelle Loiodice, Chuanhai Fu, Damian Brunner, François J Nédélec, and Phong T Tran. 2007. "Crosslinkers and Motors Organize Dynamic Microtubules to Form Stable Bipolar Arrays in Fission Yeast." *Cell* 128 (2): 357–68. doi:10.1016/j.cell.2006.12.030.

Jonsson, Erik, Moé Yamada, Ronald D Vale, and Gohta Goshima. 2015. "Clustering of a Kinesin-14 Motor Enables Processive Retrograde Microtubule-Based Transport in Plants." *Nature Plants* 1 (7): 547. Nature Publishing Group. doi:10.1038/nplants.2015.87.

Kleylein-Sohn, Julia, Bernadette Pöllinger, Michaela Ohmer, Francesco Hofmann, Erich A Nigg, Brian A Hemmings, and Markus Wartmann. 2013. "Acentrosomal Spindle Organization Renders Cancer Cells Dependent on the Kinesin HSET." *Journal of Cell Science* 125 (22): 5391–402. doi:10.1242/jcs.107474.

Kwon, M, S A Godinho, N S Chandhok, N J Ganem, A Azioune, M Thery, and D Pellman. 2008. "Mechanisms to Suppress Multipolar Divisions in Cancer Cells with Extra Centrosomes." *Genes & Development* 22 (16): 2189–203. doi:10.1101/gad.1700908.

Lansky, Zdenek, Marcus Braun, Annemarie Lüdecke, Michael Schlierf, Pieter Rein ten Wolde, Marcel E Janson, and Stefan Diez. 2015. "Diffusible Crosslinkers Generate Directed Forces in Microtubule Networks." *Cell* 160 (6): 1159–68. doi:10.1016/j.cell.2015.01.051.

Lawrence, Carolyn J, R Kelly Dawe, Karen R Christie, Don W Cleveland, Scott C Dawson, Sharyn A Endow, Lawrence S B Goldstein, et al. 2004. "A Standardized Kinesin Nomenclature: Table I." *The Journal of Cell Biology* 167 (1): 19–22. doi:10.1083/jcb.200408113.

Lecland, Nicolas, and Jens Lüders. 2014. "The Dynamics of Microtubule Minus Ends in the Human Mitotic Spindle." *Nature Cell Biology* 16 (8): 770–8. doi:10.1038/ncb2996.

Li, Yonghe, Wenyan Lu, Dongquan Chen, Rebecca J Boohaker, Ling Zhai, Indira Padmalayam, Krister Wennerberg, Bo Xu, and Wei Zhang. 2015. "KIFC1 Is a Novel Potential Therapeutic Target for Breast Cancer." *Cancer Biology & Therapy* 16 (9): 1316–22. doi:10.1080/15384047.2015.1070980.

Liu, Hong-Lei, Charles W Pemble, and Sharyn A Endow. 2012. "Neck-Motor Interactions Trigger Rotation of the Kinesin Stalk." *Scientific Reports* 2 (1): 236. Nature Publishing Group. doi:10.1038/srep00236.

Lüdecke, Annemarie, Anja-Maria Seidel, Marcus Braun, Zdenek Lansky, and Stefan Diez. 2018. "Diffusive Tail Anchorage Determines Velocity and Force Produced by Kinesin-14 Between Crosslinked Microtubules." *Nature Communications* May: 1–9. Springer US. doi:10.1038/s41467-018-04656-0.

Makino, Tsukasa, Hisayuki Morii, Takashi Shimizu, Fumio Arisaka, Yusuke Kato, Koji Nagata, and Masaru Tanokura. 2007. "Reversible and Irreversible Coiled Coils in the Stalk Domain of Ncd Motor Protein." *Biochemistry* 46 (33): 9523–32. American Chemical Society. doi:10.1021/bi700291a.

Manning, Brendan D, Jennifer G Barrett, Julie A Wallace, Howard Granok, and Michael Snyder. 1999. "Differential Regulation of the Kar3p Kinesin-Related Protein by Two Associated Proteins, Cik1p and Vik1p." *The Journal of Cell Biology* 144 (6): 1219–33. doi:10.1083/jcb.144.6.1219.

Matthies, H J, H B McDonald, L S Goldstein, and W E Theurkauf. 1996. "Anastral Meiotic Spindle Morphogenesis: Role of the Non-Claret Disjunctional Kinesin-Like Protein." *The Journal of Cell Biology* 134 (2): 455–64. doi:10.1083/jcb.134.2.455.

McDonald, H B, and L S Goldstein. 1990. "Identification and Characterization of a Gene Encoding a Kinesin-Like Protein in Drosophila." *Cell* 61 (6): 991–1000. doi:10.1016/0092-8674(90)90064-1.

McDonald, H B, R J Stewart, and L S Goldstein. 1990. "The Kinesin-Like Ncd Protein of Drosophila Is a Minus End-Directed Microtubule Motor." *Cell* 63 (6): 1159–65.

Mieck, Christine, Maxim I Molodtsov, Katarzyna Drzewicka, Babet van der Vaart, Gabriele Litos, Gerald Schmauss, Alipasha Vaziri, and Stefan Westermann. 2015. "Non-Catalytic Motor Domains Enable Processive Movement and Functional Diversification of the Kinesin-14 Kar3." *eLife* 4 (January): 1161. doi:10.7554/eLife.04489.

Mountain, V, C Simerly, L Howard, A Ando, G Schatten, and D A Compton. 1999. "The Kinesin-Related Protein, HSET, Opposes the Activity of Eg5 and Cross-Links Microtubules in the Mammalian Mitotic Spindle." *The Journal of Cell Biology* 147 (2): 351–66.

Nigg, Erich A, Lukáš Čajánek, and Christian Arquint. 2014. "The Centrosome Duplication Cycle in Health and Disease." *FEBS Letters* 588 (15): 2366–72. doi:10.1016/j.febslet.2014.06.030.

Nitzsche, Bert, Elzbieta Dudek, Lukasz Hajdo, Andrzej A Kasprzak, Andrej Vilfan, and Stefan Diez. 2016. "Working Stroke of the Kinesin-14, Ncd, Comprises Two Substeps of Different Direction." *Proceedings of the National Academy of Sciences* 113 (43): E6582–9. doi:10.1073/pnas.1525313113.

Norris, Stephen R, Seungyeon Jung, Prashant Singh, Claire E Strothman, Amanda L Erwin, Melanie D Ohi, Marija Zanic, and Ryoma Ohi. 2018. "Microtubule Minus-End Aster Organization is Driven by Processive HSET-Tubulin Clusters." *Nature Communications* 9 (1): 2659. Nature Publishing Group. doi:10.1038/s41467-018-04991-2.

Oladipo, Abiola, Ann Cowan, and Vladimir Rodionov. 2007. "Microtubule Motor Ncd Induces Sliding of Microtubules in Vivo." Edited by Ted Salmon. *Molecular Biology of the Cell* 18 (9): 3601–6. doi:10.1091/mbc.e06-12-1085.

Olmsted, Zachary T, Andrew G Colliver, Timothy D Riehlman, and Janet L Paluh. 2014. "Kinesin-14 and Kinesin-5 Antagonistically Regulate Microtubule Nucleation by Γ-TuRC in Yeast and Human Cells." *Nature Communications* 5 (1): 5339. Nature Publishing Group. doi:10.1038/ncomms6339.

Page, B D, L L Satterwhite, M D Rose, and M Snyder. 1994. "Localization of the Kar3 Kinesin Heavy Chain-Related Protein Requires the Cik1 Interacting Protein." *The Journal of Cell Biology* 124 (4): 507–19. Rockefeller University Press. doi:10.1083/jcb.124.4.507.

Pechatnikova, E, and E W Taylor. 1997. "Kinetic Mechanism of Monomeric Non-Claret Disjunctional Protein (Ncd) ATPase." *Journal of Biological Chemistry* 272 (49): 30735–40. American Society for Biochemistry and Molecular Biology. doi:10.1074/jbc.272.49.30735.

Pechatnikova, E, and E W Taylor. 1999. "Kinetics Processivity and the Direction of Motion of Ncd." *Biophysj* 77 (2): 1003–16. doi:10.1016/S0006-3495(99)76951-1.

Popchock, Andrew R, Kuo-Fu Tseng, Pan Wang, P Andrew Karplus, Xin Xiang, and Weihong Qiu. 2017. "The Mitotic Kinesin-14 KlpA Contains a Context-Dependent Directionality Switch." *Nature Communications* 8 (January): 13999. Nature Publishing Group. doi:10.1038/ncomms13999.

Rank, Katherine C, Chun Ju Chen, Julia Cope, Ken Porche, Andreas Hoenger, Susan P Gilbert, and Ivan Rayment. 2012. "Kar3Vik1, a Member of the Kinesin-14 Superfamily, Shows a Novel Kinesin Microtubule Binding Pattern." *The Journal of Cell Biology* 197 (7): 957–70. doi:10.1083/jcb.201201132.

Reddy, A S, and I S Day. 2001. "Kinesins in the Arabidopsis Genome: A Comparative Analysis Among Eukaryotes." *BMC Genomics* 2: 2. BioMed Central.

Ring, D. 1982. "Mitosis in a Cell with Multiple Centrioles." *The Journal of Cell Biology* 94 (3): 549–56. doi:10.1083/jcb.94.3.549.

Rodriguez, Adrianna S, Joseph Batac, Alison N Killilea, Jason Filopei, Dimitre R Simeonov, Ida Lin, and Janet L Paluh. 2008. "Protein Complexes at the Microtubule Organizing Center Regulate Bipolar Spindle Assembly." *Cell Cycle (Georgetown, Tex.)* 7 (9): 1246–53. Taylor & Francis. doi:10.4161/cc.7.9.5808.

Roostalu, Johanna, Jamie Rickman, Claire Thomas, François Nédélec, and Thomas Surrey. 2018. "Determinants of Polar Versus Nematic Organization in Networks of Dynamic Microtubules and Mitotic Motors." *Cell* 175 (3): 796–808.e14. doi:10.1016/j.cell.2018.09.029.

Rosenfeld, S S, B Rener, J J Correia, M S Mayo, and H C Cheung. 1996. "Equilibrium Studies of Kinesin-Nucleotide Intermediates." *Journal of Biological Chemistry* 271 (16): 9473–82. American Society for Biochemistry and Molecular Biology. doi:10.1074/jbc.271.16.9473.

Sablin, E P, R B Case, S C Dai, C L Hart, A Ruby, R D Vale, and R J Fletterick. 1998. "Direction Determination in the Minus-End-Directed Kinesin Motor Ncd." *Nature* 395 (6704): 813–6. Nature Publishing Group. doi:10.1038/27463.

Saunders, W, V Lengyel, and M A Hoyt. 1997. "Mitotic Spindle Function in Saccharomyces Cerevisiae Requires a Balance Between Different Types of Kinesin-Related Motors." *Molecular Biology of the Cell* 8 (6): 1025–33. doi:10.1091/mbc.8.6.1025.

Sharp, D J, K R Yu, J C Sisson, W Sullivan, and J M Scholey. 1999. "Antagonistic Microtubule-Sliding Motors Position Mitotic Centrosomes in Drosophila Early Embryos." *Nature Cell Biology* 1 (1): 51–4. Nature Publishing Group. doi:10.1038/9025.

Sharp, D J, H M Brown, M Kwon, G C Rogers, G Holland, and J M Scholey. 2000. "Functional Coordination of Three Mitotic Motors in Drosophila Embryos." Edited by J Richard McIntosh. *Molecular Biology of the Cell* 11 (1): 241–53. doi:10.1091/mbc.11.1.241.

She, Zhen-Yu, Meng-Ying Pan, Fu-Qing Tan, and Wan-Xi Yang. 2017. "Minus End-Directed Kinesin-14 KIFC1 Regulates the Positioning and Architecture of the Golgi Apparatus." *Oncotarget* 8 (22): 36469–83. doi:10.18632/oncotarget.16863.

Stewart, R J, J Semerjian, and C F Schmidt. 1998. "Highly Processive Motility Is Not a General Feature of the Kinesins." *European Biophysics Journal* 27 (4): 353–60.

Sturtevant, Alfred Henry. 1929. "The Claret Mutant Type of Drosophila Simulans: A Study of Chromosome Elimination and of Cell-Iineage." *Zeitschrift für wissenschaftliche Zoologie* 135: 323–56.

Suetsugu, Noriyuki, Noboru Yamada, Takatoshi Kagawa, Hisashi Yonekura, Taro Q P Uyeda, Akeo Kadota, and Masamitsu Wada. 2010. "Two Kinesin-Like Proteins Mediate Actin-Based Chloroplast Movement in Arabidopsis Thaliana." *Proceedings of the National Academy of Sciences of the United States of America* 107 (19): 8860–5. doi:10.1073/pnas.0912773107.

Syrovatkina, Viktoriya, and Phong T Tran. 2015. "Loss of Kinesin-14 Results in Aneuploidy via Kinesin-5-Dependent Microtubule Protrusions Leading to Chromosome Cut." *Nature Communications* 6 (May): 1–8. Nature Publishing Group. doi:10.1038/ncomms8322.

Szczesna, Ewa, and Andrzej A Kasprzak. 2012. "The C-Terminus of Kinesin-14 Ncd Is a Crucial Component of the Force Generating Mechanism." *FEBS Letters* 586 (6): 854–8. doi:10.1016/j.febslet.2012.02.011.

Tanaka, Kozo, Naomi Mukae, Hilary Dewar, Mark van Breugel, Euan K James, Alan R Prescott, Claude Antony, and Tomoyuki U Tanaka. 2005. "Molecular Mechanisms of Kinetochore Capture by Spindle Microtubules." *Nature* 434 (7036): 987–94. Nature Publishing Group. doi:10.1038/nature03483.

Tanaka, Kozo, Etsushi Kitamura, Yoko Kitamura, and Tomoyuki U Tanaka. 2007. "Molecular Mechanisms of Microtubule-Dependent Kinetochore Transport Toward Spindle Poles." *The Journal of Cell Biology* 178 (2): 269–81. Rockefeller University Press. doi:10.1083/jcb.200702141.

Tao, Li, Alex Mogilner, Gul Civelekoglu-Scholey, Roy Wollman, James Evans, Henning Stahlberg, and Jonathan M Scholey. 2006. "A Homotetrameric Kinesin-5, KLP61F, Bundles Microtubules and Antagonizes Ncd in Motility Assays." *Current Biology* 16 (23): 2293–302. doi:10.1016/j.cub.2006.09.064.

Tseng, Kuo-Fu, Pan Wang, Yuh-Ru Julie Lee, Joel Bowen, Allison M Gicking, Lijun Guo, Bo Liu, and Weihong Qiu. 2018. "The Preprophase Band-Associated Kinesin-14 OsKCH2 Is a Processive Minus-End-Directed Microtubule Motor." *Nature Communications* 9 (1): 1067. Nature Publishing Group. doi:10.1038/s41467-018-03480-w.

Walczak, C E, S Verma, and T J Mitchison. 1997. "XCTK2: A Kinesin-related Protein That Promotes Mitotic Spindle Assembly in *Xenopus laevis* Egg Extracts." *Journal of Cell Biology* 136: 859–70.

Walczak, C E, I Vernos, T J Mitchison, E Karsenti, and R Heald. 1998. "A Model for the Proposed Roles of Different Microtubule-Based Motor Proteins in Establishing Spindle Bipolarity." *Current Biology* 8 (16): 903–13.

Walker, R A, E D Salmon, and S A Endow. 1990. "The Drosophila Claret Segregation Protein Is a Minus-End Directed Motor Molecule." *Nature* 347 (6295): 780–2. Nature Publishing Group. doi:10.1038/347780a0.

Walter, Wilhelm J, Isabel Machens, Fereshteh Rafieian, and Stefan Diez. 2015. "The Non-Processive Rice Kinesin-14 OsKCH1 Transports Actin Filaments Along Microtubules with Two Distinct Velocities." *Nature Plants* 1 (8): 15111. Nature Publishing Group. doi:10.1038/nplants.2015.111.

Watts, Ciorsdaidh A, Frances M Richards, Andreas Bender, Peter J Bond, Oliver Korb, Oliver Kern, Michelle Riddick, et al. 2013. "Design, Synthesis, and Biological Evaluation of an Allosteric Inhibitor of HSET That Targets Cancer Cells with Supernumerary Centrosomes." *Chemistry & Biology* 20 (11): 1399–410. The Authors. doi:10.1016/j.chembiol.2013.09.012.

Weaver, Lesley N, Stephanie C Ems-McClung, Sez-Hon R Chen, Ge Yang, Sidney L Shaw, and Claire E Walczak. 2015. "The Ran-GTP Gradient Spatially Regulates XCTK2 in the Spindle." *Current Biology* 25 (11): 1509–14. Elsevier Ltd. doi:10.1016/j.cub.2015.04.015.

Wendt, Thomas G, Niels Volkmann, Georgios Skiniotis, Kenneth N Goldie, Jens Müller, Eckhard Mandelkow, and Andreas Hoenger. 2002. "Microscopic Evidence for a Minus-End-Directed Power Stroke in the Kinesin Motor Ncd." *The EMBO Journal* 21 (22): 5969–78. doi:10.1093/emboj/cdf622.

Wu, Jiaquan, Keith Mikule, Wenxian Wang, Nancy Su, Philip Petteruti, Farzin Gharahdaghi, Erin Code, et al. 2013. "Discovery and Mechanistic Study of a Small Molecule Inhibitor for Motor Protein KIFC1." *ACS Chemical Biology* 8 (10): 2201–8. doi:10.1021/cb400186w.

Yamada, Moé, Yohko Tanaka-Takiguchi, Masahito Hayashi, Momoko Nishina, and Gohta Goshima. 2017. "Multiple Kinesin-14 Family Members Drive Microtubule Minus End-Directed Transport in Plant Cells." *The Journal of Cell Biology* 216 (6): 1705–14. Rockefeller University Press. doi:10.1083/jcb.201610065.

Yamada, Moé, and Gohta Goshima. 2018. "The KCH Kinesin Drives Nuclear Transport and Cytoskeletal Coalescence to Promote Tip Cell Growth in Physcomitrella Patens." *The Plant Cell* 30 (7): 1496–510. American Society of Plant Biologists. doi:10.1105/tpc.18.00038.

Yamagishi, Masahiko, Hideki Shigematsu, Takeshi Yokoyama, Masahide Kikkawa, Mitsuhiro Sugawa, Mari Aoki, Mikako Shirouzu, Junichiro Yajima, and Ryo Nitta. 2016. "Structural Basis of Backwards Motion in Kinesin-1-Kinesin-14 Chimera: Implication for Kinesin-14 Motility." *Structure (London, England: 1993)* 24 (8): 1322–34. doi:10.1016/j.str.2016.05.021.

Yang, Bin, Michelle L Lamb, Tao Zhang, Edward J Hennessy, Gurmit Grewal, Li Sha, Mark Zambrowski, et al. 2014. "Discovery of Potent KIFC1 Inhibitors Using a Method of Integrated High-Throughput Synthesis and Screening." *Journal of Medicinal Chemistry* 57 (23): 9958–70. American Chemical Society. doi:10.1021/jm501179r.

Yang, Xue-Yong, Zi-Wei Chen, Tao Xu, Zhe Qu, Xiao-Di Pan, Xing-Hua Qin, Dong-Tao Ren, and Guo-Qin Liu. 2011. "Arabidopsis Kinesin KP1 Specifically Interacts with VDAC3, a Mitochondrial Protein, and Regulates Respiration During Seed Germination at Low Temperature." *The Plant Cell* 23 (3): 1093–106. American Society of Plant Biologists. doi:10.1105/tpc.110.082420.

Yukawa, Masashi, Chiho Ikebe, and Takashi Toda. 2015. "The Msd1–Wdr8–Pkl1 Complex Anchors Microtubule Minus Ends to Fission Yeast Spindle Pole Bodies." *The Journal of Cell Biology* 209 (4): 549–62. doi:10.1083/jcb.201412111.

Yun, Mikyung, C Eric Bronner, Cheon Gil Park, Sun Shin Cha, Hee Won Park, and Sharyn A Endow. 2003. "Rotation of the Stalk/Neck and One Head in a New Crystal Structure of the Kinesin Motor Protein, Ncd." *The EMBO Journal* 22 (20): 5382–9. doi:10.1093/emboj/cdg531.

Zhang, Wei, Ling Zhai, Yimin Wang, Rebecca J Boohaker, Wenyan Lu, Vandana V Gupta, Indira Padmalayam, et al. 2016. "Discovery of a Novel Inhibitor of Kinesin-Like Protein KIFC1." *The Biochemical Journal* 473 (8): 1027–35. Portland Press Limited. doi:10.1042/BJ20150992.

11 The Kinesin-15 Family

*Hauke Drechsler, Jaspreet Singh Grewal
and Andrew D. McAinsh*

CONTENTS

Kinesin-15s are a family of processive, plus-end-directed motors that can function as either homodimers or homotetramers. They can cross-link and bundle microtubules and play crucial roles in mitosis.

11.1 EXAMPLE FAMILY MEMBERS

Mammalian: KIF15 (KLP2)
Caenorhabditis elegans: KLP-18
Xenopus laevis: KLP2
Arabidopsis thaliana: POK1, POK2, PAKRP1
Physcomitrella patens: KINID1a, KINID1b

11.2 STRUCTURAL INFORMATION

Animal Kinesin-15s typically consist of a heavy chain containing ~1400 amino acids with a molecular weight of ~160 kDa (Boleti, Karsenti, and Vernos 1996, Sueishi, Takagi, and Yoneda 2000, Rogers et al. 2000, Buster et al. 2003). No further regulatory or accessory medium or light chains are required for motility (Drechsler et al. 2014, Drechsler and McAinsh 2016, Sturgill et al. 2014, Reinemann et al. 2017, Milic et al. 2018). The Kinesin-15 heavy chain features a short N-terminal extension (aa1–19 in human KIF15, hKIF15 from here on) preceding the N-terminal motor domain including the neck-linker (aa19–375 in hKIF15, crystal structure available at PDB ID: 4bn2 or in Klejnot et al. 2014), which is followed by a long alpha-helical stalk forming an interrupted coiled-coil (aa376–1148 in hKIF15) and finally by the C-terminal tail domain (aa1149–1388 in hKIF15) (Boleti, Karsenti, and Vernos 1996, Klejnot et al. 2014, Buster et al. 2003) (Figure 11.1). The function of the N-terminal extension is currently unknown. However, similar N-terminal extensions are involved in cargo binding (Kinesin-3 (Gruneberg et al. 2006) and Kinesin-7 (Drechsler, Tan, and Liakopoulos 2015, Roberts, Goodman, and Reck-Peterson 2014)), force generation (Kinesin-1 (Khalil et al. 2008, Hwang, Lang, and Karplus 2008)) or microtubule association (Kinesin-5 (Stock, Chu, and Hackney 2003, Britto et al. 2016) and Kinesin-7 (Drechsler, Tan, and Liakopoulos 2015)).

The simplest oligomerisation state for Kinesin-15 motors is a homodimer: both *Xenopus laevis* and *Strongylocentrotus purpuratus* (sea urchin) Kinesin-15, partially purified from meiotic egg extracts, are dimeric with a measured Stokes radius of ~10 nm and a sedimentation coefficient of ~8 S, consistent with being an elongated molecule with a native molecular weight of 334 kDa (Rogers et al. 2000, Wittmann et al. 1998). In contrast, the human Kinesin-15 (hKIF15) can also assemble into homotetramers (i.e., a dimer-of-dimers) both *in vivo* and *in vitro* (Mann, Balchand, and Wadsworth 2017, Drechsler et al. 2014, Drechsler and McAinsh 2016, Sturgill et al. 2014). The tetrameric and dimeric forms of hKIF15 co-exist in a salt-dependent equilibrium. Regardless of its source (i.e., recombinant or endogenous (Drechsler and McAinsh 2016)), hKIF15 is primarily tetrameric (~12 S) at low to physiological ionic strength (I = 75 mM to 225 mM), and primarily dimeric (~8 S) at an ionic strength above 300 mM (Drechsler and McAinsh 2016, Drechsler et al. 2014, Sturgill et al. 2014, Mann, Balchand, and Wadsworth 2017). It is important to note that the dimeric form of hKIF15 – purified at high ionic strength – appears to be auto-inhibited (Sturgill et al. 2014), whereas KIF15 tetramers are constitutively active (Drechsler and McAinsh 2016, Drechsler et al. 2014, Mann, Balchand, and Wadsworth 2017). As a result, experiments with the dimeric form are only possible if the motor is either truncated (i.e., aa1–700 (Sturgill et al. 2014, Milic et al. 2018)) or by releasing autoinhibition with an antibody targeting the C-terminus of hKIF15 (Sturgill et al. 2014). There is currently no structural data on how the two hKIF15 dimers arrange to form the tetramer, although mass spectrometry cross-linking experiments suggest a parallel tetramer, with all motor domains gathered at one end of the tetramer (Hussain, McAinsh and Jones, unpublished). This arrangement would be distinct from the arrangement of heavy chains in the – also tetrameric – Kinesin-5, which has overlapping functions with KIF15 (see below). In Kinesin-5s, two antiparallel coiled-coils form a four-helical bipolar assembly (BASS) domain (Acar et al. 2013, Scholey et al. 2014), resulting in a dumbbell structure and exposing a motor domain pair at each end of the rod-like tetramer (see Chapter 6 for details).

11.2.1 KINESIN-15 IN PLANTS

In contrast to their orthologues in animals, plant Kinesin-15s are poorly charac-terised at the molecular level. Currently, we know two different species of plant Kinesin-15 that are either slightly shorter (i.e., KINID1a/b: ~1200 amino acids, and PAKRP1/1L: ~1300 amino acids) or considerably longer (i.e., POK1: 2066 amino acids, and POK2: 2771 amino acids) than their animal counterparts (Figure 11.1). Studies on a truncated POK2 variant, $POK2_{1-589}$, suggest that no additional medium or light chains are required for plant Kinesin-15 motility *in vitro* and that dimers are the minimal functional unit (Chugh et al. 2018). Two further structural features distinguish plant Kinesin-15s: (i) the relative low abundance of extended coiled-coil stretches in their stalks (particularly in PAKRP1/1L and KINID1a/b) and (ii) the remarkably long N-terminal extension preceding the respective motor domains, ranging from 39 amino acids in KINID1b to the enormous 190 amino acids in POK2 (Figure 11.1). In the case of POK2, this N-terminal extension is required for the pro-cessive movement (Chugh et al. 2018), with conflicting reports as to whether there is a second ATP-independent microtubule-binding site (Chugh et al. 2018, Herrmann et al. 2018). The C-terminus of AtPOK1 (aa1683–2066) binds to the preprophase band (PPB) resident MAP TAN ("tangled") (Lipka et al. 2014), while the N-terminus (aa1–189) and C-terminus (aa2083–2771) of AtPOK2 interact with the plant PRC1/ASE1 orthologue MAP65-3, thereby restricting AtPOK2 localisation to the plant's phragmoplast midzone (Herrmann et al. 2018) (Figure 11.1).

11.3 FUNCTIONAL PROPERTIES

11.3.1 MOTILITY

The human KIF15 is a processive plus-end-directed motor, moving *in vitro* at 130–200 nm s^{-1} without load (Drechsler et al. 2014, Sturgill et al. 2014, Mann, Balchand, and Wadsworth 2017, McHugh et al. 2018). Under certain conditions, i.e., by increasing the assay temperature or the ionic strength, unloaded velocities have been reported to reach ~500 nm s^{-1} (Drechsler and McAinsh 2016, Milic et al. 2018). Processive runs can be interrupted by prolonged pauses (~5 s (Drechsler et al. 2014)) and the hKIF15 motor is capable of switching between processive and dif-fusive modes of movement (Drechsler et al. 2014). Single motors dwell at micro-tubule plus ends for ~20 s (Drechsler et al. 2014), leading to motor accumulation (Drechsler and McAinsh 2016). At microtubule intersections, hKIF15 motors can switch microtubule tracks with a likelihood of one in five, allowing the motor to nav-igate long distances through complex microtubule networks (Drechsler et al. 2014, Mann, Balchand, and Wadsworth 2017). Interestingly, short episodes of processive minus-end-directed movement have been detected, although this is infrequent and restricted to shorter run length and residency time compared to plus-end-directed motors (Drechsler et al. 2014, Mann, Balchand, and Wadsworth 2017). The oligo-merisation state does not appear to affect the motor's velocity (~140 nm s^{-1} for tet-ramers vs. ~190 nm s^{-1} for dimers that are activated by an antibody bound to the C-terminus). However, activated dimers exhibit a reduced run length (~0.6 μm vs.

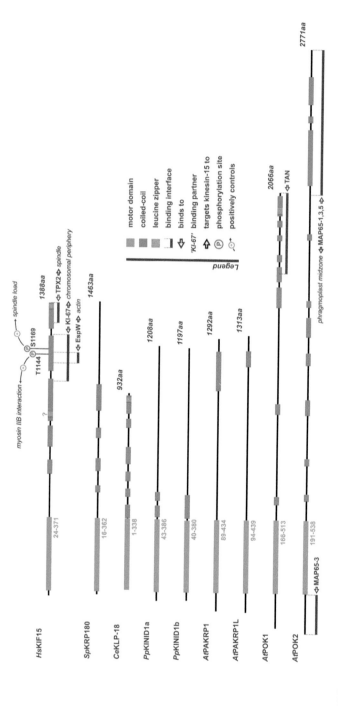

FIGURE 11.1 Scheme showing the domain organisation, the binding interfaces of interaction partners, as well as the known post-translational modifications of Kinesin-15s in animals (HsKIF15, SpKRP180, CeKLP-18) and plants (PpKINID1a/1b, AtPAKRP1/1L, AtPOK1/2). Hs – *Homo sapiens*, Sp – *Strongylocentrotus purpuratus*, Ce – *Caenorhabditis elegans*, Pp – *Physcomitrella patens*, At – *Arabidopsis thaliana*.

~2 μm for tetramers) and residency time (~4 s vs. ~26 s for tetramers) (Drechsler et al. 2014, Sturgill et al. 2014). This difference might be explained by the presence of a second pair of motor domains in the tetramer, which increases the microtubule affinity at the expense of velocity. However, the hKIF15 constructs used by Sturgill et al. (2014) lack two-thirds of the N-terminal extension, which may affect motor performance (see Structural Information Section 11.2).

11.3.2 Behaviour Under Load

During processive runs, hKIF15 motors have been reported to stall at ~3 pN (Reinemann et al. 2017, Drechsler et al. 2014) or at ~6 pN (McHugh et al. 2018, Milic et al. 2018), differences that likely depend on the construct and assay method used. External motor load strongly influences hKIF15 performance, as its velocity scales inversely with the amount of hindering forces (i.e., forces directed against the walking direction) acting on the motor (Milic et al. 2018). Furthermore, motor detachment occurs more easily under assisting loads than under hindering loads (McHugh et al. 2018) (Figure 11.2A). This implies that motors experiencing assisting forces are more prone to slip along the microtubule, while motors that experience hindering loads are more likely to grip the microtubule tightly (McHugh et al. 2018). This grip state can be further reinforced by the microtubule-associated protein (MAP) Tpx2 (targeting protein for xKlp2) (Wittmann et al. 1998, Wittmann et al. 2000), which binds to the (dimerised) leucine-zipper motif near the C-terminus of Klp2 (Wittmann et al. 1998) (aa1359–1380 in hKIF15, (Figure 11.1). In Humans, binding of TPX2 (from now on hTPX2) locks the motor even more tightly to the microtubule, allowing the motor to withstand forces (much) higher than its stall force without detaching from the microtubule (Drechsler et al. 2014, Mann, Balchand, and Wadsworth 2017) (Figure 11.2B). These hKIF15:hTPX2 complexes are likely to form *in situ* on the microtubule as, in solution, the affinity of hKIF15 and hTPX2 for one another is relatively low (Drechsler et al. 2014).

11.3.3 Microtubule Interaction

Being tetrameric, hKIF15 can cross-link and bundle microtubules via its two sets of two motor domains. However, dimeric hKIF15 also retains the capacity to cross-link microtubules via a second, nucleotide-independent microtubule-binding site in the motor domain proximal section of the stalk (aa400–700). This interaction depends on electrostatic attraction to the negatively charged tubulin C-termini exposed on the microtubule surface (Reinemann et al. 2017, Sturgill et al. 2014). The cross-linking and motor activities of recombinant full-length tetrameric hKIF15 are sufficient to sort microtubules into stable parallel bundles (Drechsler and McAinsh 2016, Reinemann et al. 2017) (Figure 11.2C). This sorting mechanism involves dynamic collectives of hKIF15 motors driving an adaptive microtubule transport/parallel sliding mechanism at microtubule intersections. While the motor domain pairs of the hKIF15 tetramer move along both microtubules which they cross-link, a velocity differential between them is evident, which gradually depends on the microtubule geometry at the intersection. On parallel microtubules, the velocity differential is small, resulting

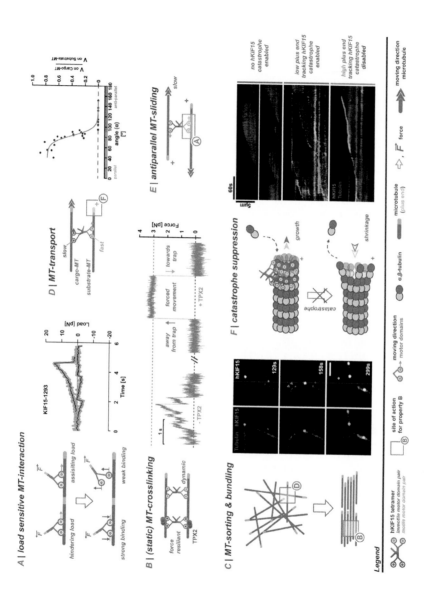

FIGURE 11.2 Functional properties of human Kinesin-15 motors based on *in vitro* data. Descriptive schemes (left) and examples of the corresponding *in vitro* data (right) are shown for each property. (A) Force-over-time profile of a C-terminally truncated (but still tetrameric) hKIF15 motor that

FIGURE 11.2 (CONTINUED)

experiences assisting (negative) or hindering (positive) loads by an optical trap. (B) Force-over-time profile of a recombinant hKIF15 motor attached to a polystyrene bead, stepping away from the centre of an optical trap in the absence of hTPX2. Force-over-time profile of the same, now stationary, bead in the presence of hTPX2 including an episode of forced movement away and towards the centre of the optical trap without detaching the motor from the microtubule. Note: the forces reached by the forced movement of the bead in the presence of hTPX2 exceed those forces generated by the motor in the absence of hTPX2. (C) Selected stills of a movie showing microtubule transport driven by collectives of recombinant hKIF15 at the intersection of dynamic microtubules, eventually aligning both microtubules in a parallel manner. Note the intensity difference between transporting hKIF15-eGFP (enhanced green fluorescent protein) collectives and single tetramers moving on the microtubule lattice. The white asterisk in the last still indicates the start position of the motors as seen in the first frame (Scale bar, 2 μm). (D) Plot showing the correlation between the velocity differential of microtubule-transporting hKIF15 motor collectives (i.e., the collective's velocity on the cargo microtubule divided by its velocity on the substrate microtubule) and the angle α between the minus ends of the cross-linked microtubules. A value of "1" means that the motor(s) move(s) on the cargo microtubule as fast as on the substrate microtubule. A value of "0" infers that the motor(s) do(es) not move at all on the cargo microtubule, although they still show motility on the substrate microtubule. (E) Schematic representation showing the suggested orientation of tetrameric hKIF15 when sliding antiparallel microtubules. One (F) Kymographs showing the dynamicity of microtubule plus ends in the absence (top) or presence of 5 nM KIF15-eGFP (middle and bottom). One example each for "low" (middle) and "high" (bottom) amounts of plus-end–tracking motors are shown. MT: microtubule. Data shown in A is originally from McHugh et al. (2018), data shown in B is from Drechsler et al. (2014) and data shown in C, D and F is obtained from Drechsler and McAinsh (2016).

in limited parallel sliding. On antiparallel microtubules, the velocity differential is high, as only one motor domain pair is moving, transporting one microtubule along the other (Drechsler and McAinsh 2016) (Figure 11.2D). Transport velocities negatively correlate with the hKIF15 motor collective size and range between 25 and 250 nm s^{-1} (Drechsler and McAinsh 2016). Using a force-calibrated optical trap, it was shown that dimeric hKIF15 supports the relative sliding of antiparallel microtubules (Reinemann et al. 2017) (Figure 11.2E). However, motor forces during antiparallel sliding are not cumulative across the overlap, and sliding occurs only at very low velocities (<1 nm s^{-1}) (Reinemann et al. 2017) compared with the velocity of single motors in an optical-trap assay (>60 nm s^{-1}) (Reinemann et al. 2017, Drechsler et al. 2014) or antiparallel microtubule sliding driven by Kinesin-5, which supports sliding velocities up to 50 nm s^{-1} (Reinemann et al. 2017, Shimamoto, Forth, and Kapoor 2015, Kapitein et al. 2005). Like other kinesins (Du, English, and Ohi 2010, Varga et al. 2009, Chen and Hancock 2015, Chen et al. 2019), hKIF15 motors additionally modulate microtubule dynamics, as plus-end-tracking hKIF15 motors suppress microtubule catastrophe events in a motor number threshold-dependent manner (Drechsler and McAinsh 2016) (Figure 11.2F).

11.4 PHYSIOLOGICAL ROLE

Kinesin-15 is a highly versatile motor that is implicated in multiple cellular processes, including mitotic spindle assembly/maintenance and post-mitotic processes such as cell migration, the axonal outgrowth of neurons or the production of red blood platelets from megakaryocytes (i.e., thrombopoiesis).

11.4.1 ROLES IN MITOSIS

Assembly and maintenance of the mitotic spindle requires the balanced activity of three mitotic motors: Kinesin-5, Kinesin-15 and dynein (van Heesbeen, Tanenbaum, and Medema 2014). Spindle assembly starts with Kinesin-5-driven separation of centrosomes in prophase, followed by nuclear envelope breakdown, at which point microtubules from the two asters interact and self-organise the bipolar spindle. During prometaphase, Kinesin-15 associates with spindle microtubules in a hTPX2-dependent manner and can compensate for loss of Kinesin-5 function (Vanneste et al. 2009, Tanenbaum et al. 2009) (Figure 11.3, upper panels). This is because Kinesin-5 and -15 both create outward, spindle-extending forces, that are counteracted by dynein-dependent inward forces (van Heesbeen, Tanenbaum, and Medema 2014). Kinesin-5 activity is essential for early spindle pole separation, as prophase and prometaphase force production by Kinesin-15 is not sufficient to counteract dynein-dependent compressing forces (Tanenbaum et al. 2009, Vanneste et al. 2009, van Heesbeen, Tanenbaum, and Medema 2014). However, once the force equilibrium has changed – either by lowering dynein-dependent forces (van Heesbeen, Tanenbaum, and Medema 2014) or by increasing Kinesin-15 activity (Sturgill and Ohi 2013, Tanenbaum et al. 2009) – Kinesin-15 is capable of driving spindle assembly even in the absence of Kinesin-5. In fact, Kinesin-15 becomes essential to spindle assembly in cells that are chronically

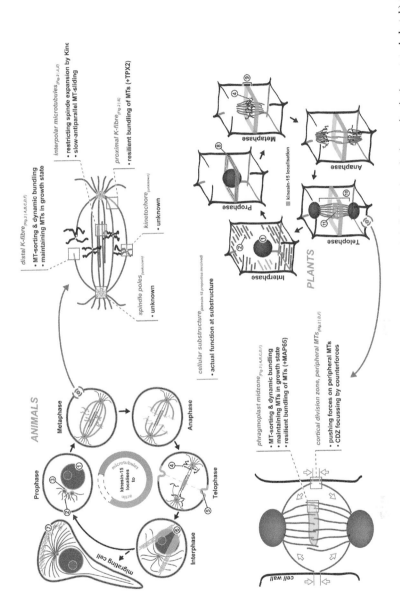

FIGURE 11.3 Localisation of Kinesin-15 motors during the cell-cycle (flowcharts) and their function during mitosis (annotated sketch) in animals (top) and plants (bottom). Please note that, due to still-missing functional *in vitro* data from plant Kinesin-15 motors, the actual functions of plant Kinesin-15s at microtubule substructures are inferred from animal Kinesin-15 data. **1** | nucleus, **2** | microtubules, **3** | microtubule-organising centre, **4** | chromosomes, **5** | cleavage furrow, **6** | actin stress fibres, **7** | lamellipodium, **8** | preprophase band, **9** | cortical division zone, **10** | phragmoplast, **11** | cell plate. Open arrows indicate forces generated by Kinesin-15 motors. MT: microtubule.

deprived of Kinesin-5 (Raaijmakers et al. 2012, Sturgill et al. 2016), and cells often increase their Kinesin-15 expression in order to maintain their capacity to divide in the absence of Kinesin-5 (Sturgill et al. 2016, Sturgill and Ohi 2013, Raaijmakers et al. 2012).

While there is a broad agreement on the mechanism by which Kinesin-5 drives spindle assembly (Kapoor 2017), the Kinesin-15 mechanism(s) remain under debate. One model proposes that Kinesin-15 – in analogy to Kinesin-5 – directly drives spindle extension by sliding apart antiparallel overlaps of interpolar microtubules at the spindle midzone (Tanenbaum et al. 2009, Sturgill et al. 2014, Reinemann et al. 2017). Experimental support for this model currently appears weak as: (i) *in vivo* hKIF15 mainly localises to k-fibres instead of localising to antiparallel microtubule overlaps in the spindle midzone (Sturgill and Ohi 2013); (ii) hKIF15 restricts rather than supports the force generation by Kinesin-5 on interpolar microtubule overlaps (Sturgill and Ohi 2013); (iii) spindle assembly by overexpressed hKIF15 in the absence of Kinesin-5 occurs late in prometaphase by a "reverse jack-knife" mechanism (Figure 11.4) (Sturgill and Ohi 2013, Toso et al. 2009). This mechanism is fundamentally different from the axial spindle elongation driven by Kinesin-5- dependent antiparallel microtubule sliding at interpolar microtubule overlaps (Kapitein et al. 2005); and (iv) antiparallel sliding of microtubules *in vitro* has not been observed with tetrameric hKIF15 motors (Drechsler and McAinsh 2016), and the sliding velocities with dimeric KIF15s (<1 nm s⁻¹) (Reinemann et al. 2017) are almost two orders of magnitude slower than those generated by ensembles of Kinesin-5 in a comparable setup (Shimamoto, Forth, and Kapoor 2015). As a consequence, assembly of a 10-μm spindle, via a linear sliding

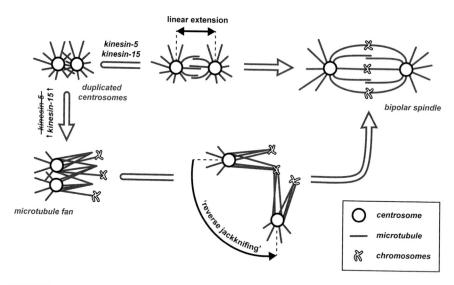

FIGURE 11.4 Scheme showing mitotic spindle assembly in the presence (top, linear extension) and absence (bottom, reverse jack-knifing) of Kinesin-5. Please note that Kinesin-15-driven spindle assembly requires additional microtubule-bundling activity e.g., by Kinesin-15 overexpression.

mechanism, would take hours when carried out by Kinesin-15, instead of taking minutes in the case of Kinesin-5. In cells, however, the contribution of Kinesin-15 to metaphase spindle elongation appears to have a magnitude of about 1 μm min^{-1} (Tanenbaum et al. 2009). An alternative model suggests that Kinesin-15 localises to k-fibres (Sturgill and Ohi 2013), where it indirectly drives spindle expansion by organising and stabilising k-fibres (Brouwers, Mallol Martinez, and Vernos 2017, van Heesbeen, Tanenbaum, and Medema 2014) through stable cross-links (Drechsler et al. 2014), while also harmonising their dynamics (Drechsler and McAinsh 2016). The idea is that k-fibres continuously grow into the attached kinetochores, creating expansive forces within the spindle that lead to centrosome separation (Toso et al. 2009). Since this mechanism requires extra bundling activity, it might explain why a cell chronically deprived of Kinesin-5 activity has to either increase hKIF15 levels (Sturgill and Ohi 2013), or provide extra bundling activity in the form of a rigor Kinesin-5 to allow spindle assembly at normal hKIF15 levels (Sturgill et al. 2016).

Finally, Kinesin-15-dependent microtubule sorting and bundling, but not antiparallel sliding, is key to microtubule rearrangements in other organisms: During the acentrosomal spindle assembly of *C. elegans* meiosis, the Kinesin-15, KLP-18, sorts spindle microtubules into parallel arrays, which enables the subsequent focusing of microtubule minus ends into defined spindle poles and therefore allows the establishment of a bipolar spindle (Wolff et al. 2016, Segbert et al. 2003). Moreover, in the moss *P. patens*, the Kinesin-15 motors, KINID1a and KINID1b, dynamically focus microtubule plus ends and bundle microtubules in apical domes of caulonemal apical cells during polarised tip growth (Hiwatashi, Sato, and Doonan 2014).

During mitosis, Kinesin-15 localises to additional cellular substructures, including the spindle poles (Buster et al. 2003, Mann, Balchand, and Wadsworth 2017, Boleti, Karsenti, and Vernos 1996), the spindle midbody (Buster et al. 2003, Rogers et al. 2000) and the kinetochores (Mann, Balchand, and Wadsworth 2017). Through an interaction of amino acids 1017–1238 with the chromosomal periphery marker KI-67 (Figure 11.1 (Booth et al. 2014, Sueishi, Takagi, and Yoneda 2000, Vanneste et al. 2009)), Kinesin-15 also localises to the chromosomal periphery (Figure 11.3, upper panels (Buster et al. 2003, Mann, Balchand, and Wadsworth 2017, Vanneste et al. 2009, Brouwers, Mallol Martinez, and Vernos 2017)). Further Kinesin-15 subpopulations have been reported to localise to the cleavage furrow during cytokinesis, as well as to actin stress fibres of interphase cells (Figure 11.3, upper panels (Buster et al. 2003)). The detection of those minor Kinesin-15 populations, however, appears to depend on the experimental conditions (i.e., visualisation method, antibodies, spatial resolution; for example, compare (Mann, Balchand, and Wadsworth 2017) with (Brouwers, Mallol Martinez, and Vernos 2017) and (Sturgill and Ohi 2013). Particular subpopulations might also be organism-specific, as Kinesin-15 is present at the spindle poles in rat (Buster et al. 2003) and frog (Boleti, Karsenti, and Vernos 1996), but not in sea urchin (Rogers et al. 2000). Conversely, Kinesin-15 localises to the spindle midbody in rat (Buster et al. 2003) and sea urchin (Rogers et al. 2000), but not in the frog (Boleti, Karsenti, and Vernos 1996). The functions of Kinesin-15 at these diverse locations are currently not known and warrant further study.

11.4.2 ESTABLISHING THE CORTICAL DIVISION ZONE AND PHRAGMOPLAST ASSEMBLY IN PLANTS

In plants, the site of cell division is already set at the beginning of mitosis. From late S-phase to prometaphase, the preprophase band (PPB) – made of endomembrane, F-actin and microtubules – lines the cell cortex and marks (but not necessarily determines) the future division site (Müller 2019). Subsequently, additional factors like the microtubule-associated protein TAN (Walker et al. 2007) and the small GTPase-activating proteins PHGAP1,2 (Stockle et al. 2016) and RANGAP1 (Xu et al. 2008) are recruited to the PPB-site in a Kinesin-15 (i.e., POK1 and POK2)-dependent manner, and persist there throughout mitosis, conserving the spatial information (now termed cortical division zone, CDZ) of the PPB after its disassembly at the end of prophase (Figure 11.3, lower panels). Targeting and tethering of POK1 and POK2 to the PPB/CDZ is mediated by their C-termini and does not depend on their motor activity or the presence of microtubules. Additionally, the C-termini of POK1 and POK2 also mediate the long-term tethering of CDZ factors at the PPB/CDZ (Figure 11.1 (Lipka et al. 2014, Herrmann et al. 2018)).

Later, during cytokinesis, plant cells divide by the radial insertion of a cell plate into the plane marked by the CDZ (Figure 11.3, lower panels). Cell plate growth is supported by the so-called phragmoplast, two arrays of parallel microtubules (i.e., remnants of the former spindle) in between the re-forming nuclei that orient with their plus ends towards the cell division plane, forming antiparallel overlaps (Ho et al. 2011, Jurgens 2005). The dynamic phragmoplast targets vesicles containing cell-plate material to the nascent cell plate and provides structural support (van Oostende-Triplet et al. 2017, Smertenko et al. 2018, Jurgens 2005). The various plant Kinesin-15 members contribute on multiple levels to the organisation of the phragmoplast. In the moss *P. patens*, the Kinesin-15s KINID1a and 1b localise to the spindle midzone from metaphase onwards and organise the antiparallel microtubule overlaps in the phragmoplast midzone during cytokinesis, keeping the phragmoplast halves together (Hiwatashi et al. 2008). Similarly, AtPAKRP1 and AtPAKRP1L localise to the spindle midzone from anaphase onwards (Lee and Liu 2000, Pan, Lee, and Liu 2004) and are believed to organise the phragmoplast midzone, as AtPAKRP1 and AtPAKRP1L double mutants fail to assemble antiparallel microtubule arrays or cell plates (Lee, Li, and Liu 2007). Also, AtPOK2 (Herrmann et al. 2018), but not AtPOK1 (Lipka et al. 2014), localises to the phragmoplast midzone. AtPOK2 mutants however show no phragmoplast assembly phenotype, but show reduced phragmoplast expansion growth (within the cell division plane) during radial cell plate growth (Herrmann et al. 2018). The mechanism by which POK2 contributes to phragmoplast expansion has not been addressed in detail yet, but it has been suggested to promote microtubule growth at the midzone – in analogy to its animal counterpart (Herrmann et al. 2018). Phragmoplast targeting of Kinesin-15 depends on its motor activity, but sub-targeting to the phragmoplast midzone depends on MAP65 isoforms that sequester AtPOK2 there via dual interactions with both the N-terminal extension and the C-terminus of POK2 (Figure 11.1 (Herrmann et al. 2018)). Finally, AtPOK1 and AtPOK2 keep the phragmoplast perpendicular to the division plane as it is marked by the cortical division zone (Müller, Han, and Smith 2006, Herrmann et al. 2018). Consequently, unaligned phragmoplasts in the absence of POK activity cause random insertion of cell walls and therefore heavily disordered plant tissues (Müller, Han, and Smith 2006, Herrmann et al. 2018). Phragmoplast alignment was suggested to occur by cortical POK motors at the CDZ, that act on peripheral phragmoplast microtubules (Chugh et al. 2018, Müller 2019). By walking to their plus ends, tethered motors create pushing forces against the expanding phragmoplast, thereby aligning it perpendicular to the CDZ plane. Simultaneously, these motors experience a counterforce by the phragmoplast which focuses them – and thereby the CDZ – on the cortex, a process that has been observed *in vivo* during progression of cytokinesis (Figure 11.3, lower left panel (Müller 2019)). Interpretation of POK1 and POK2 knockdown data, however, might be difficult, given that the CDZ identity (i.e., the localisation of TAN/PHGAP1,2/RANGAP1) strongly depends on POK-dependent targeting and tethering.

11.4.3 Postmitotic Functions in Neuronal Development

Expression of Kinesin-15 is particularly high in tissues of the vertebrate nervous system (i.e., brain, spinal cord, olfactory bulb, otic vesicle and retina) during embryonal development and early postnatal phases (Xu et al. 2014, Liu et al. 2010). Here, Kinesin-15 contributes, by as yet largely unknown mechanisms, to the control of cell morphology and cell migration. Upon depletion of Kinesin-15, neurites (i.e., dendrites and axons) are growing out faster and longer than in unperturbed control neurons, but exhibit a smaller diameter (Lin et al. 2012, Liu et al. 2010, Dong et al. 2019). This phenotype was explained by the mobilisation of short microtubule fragments in the absence of Kinesin-15 (Liu et al. 2010) and suggests that Kinesin-15 in neurons also fulfils a primarily structural, microtubule-organising function. In line with that, Kinesin-15-depleted neurons show additional microtubule-organising defects, as microtubules that orient with their plus end to the cell body are largely absent from the dendrites (Lin et al. 2012) and axons develop fewer branches and filopodia (Liu et al. 2010, Dong et al. 2019). The growth cones of Kinesin-15-depleted cells are considerably smaller than those of control cells and have lost the ability to control their growth direction (Liu et al. 2010). This phenotype might result from perturbed, Kinesin-15-dependent interactions between the microtubule and the actin cytoskeleton in the growth cone (Liu et al. 2010). In accordance, the existence of a Kinesin-15 and myosin IIB heterotetramer has been proposed, based on observations in migrating astrocytes (Feng et al. 2016). Here, disruption of the reported physical interaction between Kinesin-15 and myosin IIB (see Figure 11.1) increases astrocyte migration, phenocopying a Kinesin-15-depletion phenotype (Feng et al. 2016). In developing neurons, therefore, Kinesin-15 shows both microtubule- and actin-related activity at the same time. This stands in stark contrast to mitosis, where both functions are temporarily separated, suggesting that Kinesin-15 activities are post-translationally fine-tuned to match the respective scenarios.

11.5 INVOLVEMENT IN DISEASE

A number of studies propose hKIF15 (i.e., hKIF15 overexpression) as a biomarker for a multitude of cancer types (Liu et al. 2018, Zhao et al. 2019, Song et al. 2018, Zou et al. 2014, Sheng, Jiang, and Xue 2019, Menyhart, Pongor, and Gyorffy 2019,

Stangeland et al. 2015, Chen et al. 2017, Qiao et al. 2018, Yu et al. 2019, Wang et al. 2017), and suggest that KIF15 promotes tumour cell proliferation (Zhao et al. 2019, Zou et al. 2014, Qiao et al. 2018, Yu et al. 2019, Wang et al. 2017). However, whether hKIF15 overexpression simply reflects higher proliferative rates remains unclear and the mechanism by which hKIF15 promotes cell proliferation has not yet been established. On the other hand, there is an emerging view that hKIF15 is a valuable target for anti-cancer therapies (Rath and Kozielski 2012). While robust spindle assembly and maintenance due to the functional redundancy of Kinesins-5 and -15 is, without doubt, beneficial to the cell, this failsafe mechanism poses a problem in current anti-cancer therapy strategies. Based on the observation that loss of Kinesin-5-dependent forces completely prevents spindle assembly in most systems (Hagan and Yanagida 1992, Heck et al. 1993, Kapoor et al. 2000) and the rationale that cancer cells exhibit a higher mitotic activity compared with somatic cells, some effort has been made to develop Kinesin-5 inhibitors. These drugs, however, failed in Phase II clinical trials, presumably due to the functional redundancy of Kinesin-5 and -15 (Rath and Kozielski 2012, Chandrasekaran, Tátrai, and Gergely 2015). In fact, cells that have been chronically exposed to Kinesin-5 inhibitors acquire drug resistance in a strictly Kinesin-15-dependent manner (Sturgill et al. 2016, Sturgill and Ohi 2013). Hence, first approaches have been made to develop Kinesin-15 inhibitors that are aimed at inhibiting tumour growth when co-administered with established Kinesin-5 inhibitors (Sebastian 2017, Dumas et al. 2019, Milic et al. 2018). So far, two Kinesin-15 inhibitors have been described, KIF15-IN-1 and GW108X. KIF15-IN-1 is an ATP-competitive inhibitor (Dumas et al. 2019) that inhibits (C-terminally truncated) hKIF15 with a half-maximal inhibitory concentration (IC_{50}) of 0.2 μM (Dumas et al. 2019) to 1.7 μM (Milic et al. 2018) *in vitro*. GW108X is an allosteric inhibitor that inhibits (C-terminally truncated) hKIF15 with an IC_{50} of 0.8 μM (Dumas et al. 2019). Whereas KIF15-IN-1 is "specific" for hKIF15 (Milic et al. 2018), GW108X shows some inhibitory potential for Kinesin-5 as well (Dumas et al. 2019). Using KIF15-IN-1, it could be shown that co-administration of Kinesin-5 and -15 inhibitors inhibits the growth of cancer cells (Milic et al. 2018). Hence, these early inhibitors will be promising starting points for the development of further refined Kinesin-15 inhibitors.

A recent case study reported a mutation within the hKIF15 gene that phenocopies Braddock-Carey syndrome (BCS) (Sleiman et al. 2017). BCS is characterised by multiple developmental aberrations, mainly affecting the morphology of the head (i.e., Pierre-Robin sequence and distinctive facies) and the development of the brain (i.e., microcephaly and agenesis of the corpus callosum), as well as by a congenital thrombocytopaenia (Braddock et al. 2016). Classically, BCS is caused by microdeletions on chromosome 21 (i.e., 21q22.11), affecting at least three different contiguous genes (*SON*, *ITSN1* and *RUNX1*) (Braddock et al. 2016), each of which could be linked to certain phenotypes of BCS. Such deletions affecting the haematopoietic transcription factor *RUNX1* (Okuda et al. 2001, Sood, Kamikubo, and Liu 2017), are associated with the occurrence of congenital thrombocytopenia. In this light, it is even more surprising that a single mutation within the hKIF15 open reading frame, introducing a premature stop codon (R501*), can phenocopy such a complex syndrome (Sleiman et al. 2017). However, it emphasises the role of Kinesin-15 during neuronal development and points to a – so far unknown – role of Kinesin-15 in thrombopoiesis, a process that relies heavily on a functional actin and microtubule cytoskeleton (Favier and Raslova 2015).

REFERENCES

Acar, S., D. B. Carlson, M. S. Budamagunta, V. Yarov-Yarovoy, J. J. Correia, M. R. Ninonuevo, W. T. Jia, L. Tao, J. A. Leary, J. C. Voss, J. E. Evans, and J. M. Scholey. 2013. "The bipolar assembly domain of the mitotic motor kinesin-5." *Nature Communications* 4. doi:ARTN 134310.1038/ncomms2348.

Boleti, H., E. Karsenti, and I. Vernos. 1996. "Xklp2, a novel Xenopus centrosomal kinesin-like protein required for centrosome separation during mitosis." *Cell* 84 (1):49–59. doi:S0092-8674(00)80992-7 [pii].

Booth, D. G., M. Takagi, L. Sanchez-Pulido, E. Petfalski, G. Vargiu, K. Samejima, N. Imamoto, C. P. Ponting, D. Tollervey, W. C. Earnshaw, and P. Vagnarelli. 2014. "Ki-67 is a PP1-interacting protein that organises the mitotic chromosome periphery." *Elife* 3:e01641. doi:10.7554/eLife.01641.

Braddock, S. R., S. T. South, J. D. Schiffman, M. Longhurst, L. R. Rowe, and J. C. Carey. 2016. "Braddock-Carey syndrome: A 21q22 contiguous gene syndrome encompassing RUNX1." *American Journal of Medical Genetics Part A* 170 (10):2580–6. doi:10.1002/ajmg.a.37870.

Britto, M., A. Goulet, S. Rizvi, O. von Loeffelholz, C. A. Moores, and R. A. Cross. 2016. "Schizosaccharomyces pombe kinesin-5 switches direction using a steric blocking mechanism." *Proceedings of the National Academy of Sciences of the United States of America* 113 (47):E7483–9. doi:10.1073/pnas.1611581113.

Brouwers, Nathalie, Nuria Mallol Martinez, and Isabelle Vernos. 2017. "Role of Kif15 and its novel mitotic partner KBP in K-fiber dynamics and chromosome alignment." *PLoS one* 12 (4):e0174819. doi:10.1371/journal.pone.0174819.

Buster, Daniel W., Douglas H. Baird, Wenqian Yu, Joanna M. Solowska, Muriel Chauvière, Agnieszka Mazurek, Michel Kress, and Peter W. Baas. 2003. "Expression of the mitotic kinesin Kif15 in postmitotic neurons: Implications for neuronal migration and development." *Journal of Neurocytology* 32 (1):79–96. doi:10.1023/a:1027332432740.

Chandrasekaran, Gayathri, Péter Tátrai, and Fanni Gergely. 2015. "Hitting the brakes: Targeting microtubule motors in cancer." *British Journal of Cancer* 113:693. doi:10.1038/bjc.2015.264.

Chen, G. Y., J. M. Cleary, A. B. Asenjo, Y. Chen, J. A. Mascaro, D. F. J. Arginteanu, H. Sosa, and W. O. Hancock. 2019. "Kinesin-5 promotes microtubule nucleation and assembly by stabilizing a lattice-competent conformation of tubulin." *Current Biology* 29 (14):2259–69 e4. doi:10.1016/j.cub.2019.05.075.

Chen, J., S. Li, S. Zhou, S. Cao, Y. Lou, H. Shen, J. Yin, and G. Li. 2017. "Kinesin superfamily protein expression and its association with progression and prognosis in hepatocellular carcinoma." *Journal of Cancer Research and Therapeutics* 13 (4):651–9. doi:10.4103/jcrt.JCRT_491_17.

Chen, Y. L., and W. O. Hancock. 2015. "Kinesin-5 is a microtubule polymerase." *Nature Communications* 6. doi:ARTN 816010.1038/ncomms9160.

Chugh, M., M. Reissner, M. Bugiel, E. Lipka, A. Herrmann, B. Roy, S. Muller, and E. Schaffer. 2018. "Phragmoplast orienting kinesin 2 is a weak motor switching between processive and diffusive modes." *Biophysical Journal* 115 (2):375–85. doi:10.1016/j.bpj.2018.06.012.

Dong, Zhangji, Shuwen Wu, Chenwen Zhu, Xueting Wang, Yuanyuan Li, Xu Chen, Dong Liu, Liang Qiang, Peter W. Baas, and Mei Liu. 2019. "Clustered Regularly Interspaced Short Palindromic Repeats (CRISPR)/Cas9-mediated kif15 mutations accelerate axonal outgrowth during neuronal development and regeneration in zebrafish." *Traffic* 20 (1):71–81. doi:10.1111/tra.12621.

Drechsler, H., T. McHugh, M. R. Singleton, N. J. Carter, and A. D. McAinsh. 2014. "The Kinesin-12 Kif15 is a processive track-switching tetramer." *Elife* 3:e01724. doi:10.7554/eLife.01724.

Drechsler, H., A. N. Tan, and D. Liakopoulos. 2015. "Yeast GSK-3 kinase regulates astral microtubule function through phosphorylation of the microtubule-stabilizing kinesin Kip2." *Journal of Cell Science* 128 (21):3910–21. doi:10.1242/jcs.166686.

Drechsler, Hauke, and Andrew D. McAinsh. 2016. "Kinesin-12 motors cooperate to suppress microtubule catastrophes and drive the formation of parallel microtubule bundles." *Proceedings of the National Academy of Sciences of the United States of America* 113 (12):E1635–44. doi:10.1073/pnas.1516370113.

Du, Y., C. A. English, and R. Ohi. 2010. "The kinesin-8 Kif18A dampens microtubule plus-end dynamics." *Current Biology* 20 (4):374–80. doi:10.1016/j.cub.2009.12.049.

Dumas, Megan E., Geng-Yuan Chen, Nicole D. Kendrick, George Xu, Scott D. Larsen, Somnath Jana, Alex G. Waterson, Joshua A. Bauer, William Hancock, Gary A. Sulikowski, and Ryoma Ohi. 2019. "Dual inhibition of Kif15 by oxindole and quinazolinedione chemical probes." *Bioorganic & Medicinal Chemistry Letters* 29 (2):148–54. doi:10.1016/j.bmcl.2018.12.008.

Favier, R., and H. Raslova. 2015. "Progress in understanding the diagnosis and molecular genetics of macrothrombocytopenias." *British Journal of Haematology* 170 (5):626–39. doi:10.1111/bjh.13478.

Feng, Jie, Zunlu Hu, Haijiao Chen, Juan Hua, Ronghua Wu, Zhangji Dong, Liang Qiang, Yan Liu, Peter W. Baas, and Mei Liu. 2016. "Depletion of kinesin-12, a myosin-IIB-interacting protein, promotes migration of cortical astrocytes." *Journal of Cell Science* 129 (12):2438–47. doi:10.1242/jcs.181867.

Gruneberg, U., R. Neef, X. Li, E. H. Chan, R. B. Chalamalasetty, E. A. Nigg, and F. A. Barr. 2006. "KIF14 and citron kinase act together to promote efficient cytokinesis." *The Journal of Cell Biology* 172 (3):363–72. doi:10.1083/jcb.200511061.

Hagan, Iain, and Mitsuhiro Yanagida. 1992. "Kinesin-related cut 7 protein associates with mitotic and meiotic spindles in fission yeast." *Nature* 356 (6364):74–6. doi:10.1038/356074a0.

Heck, M M, A Pereira, P Pesavento, Y Yannoni, A C Spradling, and L S Goldstein. 1993. "The kinesin-like protein KLP61F is essential for mitosis in Drosophila." *The Journal of Cell Biology* 123 (3):665–79. doi:10.1083/jcb.123.3.665.

Herrmann, A., P. Livanos, E. Lipka, A. Gadeyne, M. T. Hauser, D. Van Damme, and S. Muller. 2018. "Dual localized kinesin-12 POK2 plays multiple roles during cell division and interacts with MAP65-3." *Embo Reports* 19 (9). doi:ARTN e4608510.15252/embr.201846085.

Hiwatashi, Y., M. Obara, Y. Sato, T. Fujita, T. Murata, and M. Hasebe. 2008. "Kinesins are indispensable for interdigitation of phragmoplast microtubules in the moss Physcomitrella patens." *Plant Cell* 20 (11):3094–106. doi:10.1105/tpc.108.061705.

Hiwatashi, Y., Y. Sato, and J. H. Doonan. 2014. "Kinesins have a dual function in organizing microtubules during both tip growth and cytokinesis in physcomitrella patens." *Plant Cell* 26 (3):1256–66. doi:10.1105/tpc.113.121723.

Ho, C. M. K., T. Hotta, F. L. Guo, R. Roberson, Y. R. J. Lee, and B. Liu. 2011. "Interaction of antiparallel microtubules in the phragmoplast is mediated by the microtubule-associated protein MAP65-3 in Arabidopsis." *Plant Cell* 23 (8):2909–23. doi:10.1105/tpc.110.078204.

Hwang, Wonmuk, Matthew J. Lang, and Martin Karplus. 2008. "Force generation in kinesin hinges on cover-neck bundle formation." *Structure* 16 (1):62–71. doi:10.1016/j.str.2007.11.008.

Jurgens, G. 2005. "Cytokinesis in higher plants." *Annual Review of Plant Biology* 56:281–99. doi:10.1146/annurev.arplant.55.031903.141636.

Kapitein, L. C., E. J. G. Peterman, B. H. Kwok, J. H. Kim, T. M. Kapoor, and C. F. Schmidt. 2005. "The bipolar mitotic kinesin Eg5 moves on both microtubules that it crosslinks." *Nature* 435 (7038):114–8. doi:10.1038/nature03503.

Kapoor, Tarun M. 2017. "Metaphase spindle assembly." *Biology* 6 (1):8.

Kapoor, Tarun M., Thomas U. Mayer, Margaret L. Coughlin, and Timothy J. Mitchison. 2000. "Probing spindle assembly mechanisms with monastrol, a small molecule inhibitor of the mitotic kinesin, Eg5." *The Journal of Cell Biology* 150 (5):975–88. doi:10.1083/jcb.150.5.975.

Khalil, Ahmad S., David C. Appleyard, Anna K. Labno, Adrien Georges, Martin Karplus, Angela M. Belcher, Wonmuk Hwang, and Matthew J. Lang. 2008. "Kinesin's cover-neck bundle folds forward to generate force." *Proceedings of the National Academy of Sciences* 105 (49):19247–52. doi:10.1073/pnas.0805147105.

Klejnot, M., A. Falnikar, V. Ulaganathan, R. A. Cross, P. W. Baas, and F. Kozielski. 2014. "The crystal structure and biochemical characterization of Kif15: A bifunctional molecular motor involved in bipolar spindle formation and neuronal development." *Acta Crystallographica Section D: Biological Crystallography* 70 (Pt 1):123–33. doi:10.1107/S1399004713028721.

Lee, Y. R., Y. Li, and B. Liu. 2007. "Two Arabidopsis phragmoplast-associated kinesins play a critical role in cytokinesis during male gametogenesis." *Plant Cell* 19 (8):2595–605. doi:10.1105/tpc.107.050716.

Lee, Y. R., and B. Liu. 2000. "Identification of a phragmoplast-associated kinesin-related protein in higher plants." *Current Biology* 10 (13):797–800.

Lin, Shen, Mei Liu, Olga I. Mozgova, Wenqian Yu, and Peter W. Baas. 2012. "Mitotic motors coregulate microtubule patterns in axons and dendrites." *The Journal of Neuroscience* 32 (40):14033–49. doi:10.1523/jneurosci.3070-12.2012.

Lipka, Elisabeth, Astrid Gadeyne, Dorothee Stöckle, Steffi Zimmermann, Geert De Jaeger, David W. Ehrhardt, Viktor Kirik, Daniel Van Damme, and Sabine Müller. 2014. "The phragmoplast-orienting kinesin-12 class proteins translate the positional information of the preprophase band to establish the cortical division zone in Arabidopsis thaliana." *Plant Cell* 26 (6):2617–32. doi:10.1105/tpc.114.124933.

Liu, M., V. C. Nadar, F. Kozielski, M. Kozlowska, W. Yu, and P. W. Baas. 2010. "Kinesin-12, a mitotic microtubule-associated motor protein, impacts axonal growth, navigation, and branching." *The Journal of Neuroscience* 30 (44):14896–906. doi:10.1523/JNEUROSCI.3739-10.2010.

Liu, M., Y. L. Qiu, T. Yin, Y. Zhou, Z. Y. Mao, and Y. J. Zhang. 2018. "Meta-analysis of microarray datasets identify several chromosome segregation- related cancer/testis genes potentially contributing to anaplastic thyroid carcinoma." *PeerJ* 6. doi:ARTN e582210.7717/peerj.5822.

Mann, B. J., S. K. Balchand, and P. Wadsworth. 2017. "Regulation of Kif15 localization and motility by the C-terminus of TPX2 and microtubule dynamics." *Molecular Biology of the Cell* 28 (1):65–75. doi:10.1091/mbc.E16-06-0476.

McHugh, T., H. Drechsler, A. D. McAinsh, N. J. Carter, and R. A. Cross. 2018. "Kif15 functions as an active mechanical ratchet." *Molecular Biology of the Cell* 29 (14):1743–52. doi:10.1091/mbc.E18-03-0151.

Menyhart, O., L. S. Pongor, and B. Gyorffy. 2019. "Mutations defining patient cohorts with elevated PD-L1 expression in gastric cancer." *Frontiers in Pharmacology* 9. doi:ARTN 152210.3389/fphar.2018.01522.

Milic, Bojan, Anirban Chakraborty, Kyuho Han, Michael C. Bassik, and Steven M. Block. 2018. "KIF15 nanomechanics and kinesin inhibitors, with implications for cancer chemotherapeutics." *Proceedings of the National Academy of Sciences* 115 (20):E4613-E4622. doi:10.1073/pnas.1801242115.

Müller, Sabine. 2019. "Plant cell division — defining and finding the sweet spot for cell plate insertion." *Current Opinion in Cell Biology* 60:9–18. doi:10.1016/j.ceb.2019.03.006.

Müller, Sabine, Shengcheng Han, and Laurie G. Smith. 2006. "Two kinesins are involved in the apatial control of cytokinesis in Arabidopsis thaliana." *Current Biology* 16 (9):888–94. doi:10.1016/j.cub.2006.03.034.

Okuda, T., M. Nishimura, M. Nakao, and Y. Fujita. 2001. "RUNX1/AML1: A central player in hematopoiesis." *International Journal of Hematology* 74 (3):252–7.

Pan, R., Y. R. Lee, and B. Liu. 2004. "Localization of two homologous Arabidopsis kinesin-related proteins in the phragmoplast." *Planta* 220 (1):156–64. doi:10.1007/s00425-004-1324-4.

Qiao, Y., J. Chen, C. Ma, Y. Liu, P. Li, Y. Wang, L. Hou, and Z. Liu. 2018. "Increased KIF15 expression predicts a poor prognosis in patients with lung adenocarcinoma." *Cellular Physiology and Biochemistry* 51 (1):1–10. doi:10.1159/000495155.

Raaijmakers, J. A., R. G. van Heesbeen, J. L. Meaders, E. F. Geers, B. Fernandez-Garcia, R. H. Medema, and M. E. Tanenbaum. 2012. "Nuclear envelope-associated dynein drives prophase centrosome separation and enables Eg5-independent bipolar spindle formation." *The EMBO Journal* 31 (21):4179–90. doi:10.1038/emboj.2012.272.

Rath, O., and F. Kozielski. 2012. "Kinesins and cancer." *Nature Reviews Cancer* 12 (8):527–39. doi:10.1038/nrc3310.

Reinemann, Dana N., Emma G. Sturgill, Dibyendu Kumar Das, Miriam Steiner Degen, Zsuzsanna Vörös, Wonmuk Hwang, Ryoma Ohi, and Matthew J. Lang. 2017. "Collective force regulation in anti-parallel microtubule gliding by dimeric Kif15 kinesin motors." *Current Biology* 27 (18):2810–20.e6. doi:10.1016/j.cub.2017.08.018.

Roberts, A. J., B. S. Goodman, and S. L. Reck-Peterson. 2014. "Reconstitution of dynein transport to the microtubule plus end by kinesin." *Elife* 3:e02641. doi:10.7554/eLife.02641.

Rogers, G. C., K. K. Chui, E. W. Lee, K. P. Wedaman, D. J. Sharp, G. Holland, R. L. Morris, and J. M. Scholey. 2000. "A kinesin-related protein, KRP(180), positions prometaphase spindle poles during early sea urchin embryonic cell division." *Journal of Cell Biology* 150 (3):499–512.

Scholey, J. E., S. Nithianantham, J. M. Scholey, and J. Al-Bassam. 2014. "Structural basis for the assembly of the mitotic motor Kinesin-5 into bipolar tetramers." *Elife* 3. doi:ARTN e0221710.7554/eLife.02217.

Sebastian, J. 2017. "Dihydropyrazole and dihydropyrrole structures based design of Kif15 inhibitors as novel therapeutic agents for cancer." *Computational Biology and Chemistry* 68:164–74. doi:10.1016/j.compbiolchem.2017.03.006.

Segbert, Christoph, Rosemarie Barkus, Jim Powers, Susan Strome, William M. Saxton, and Olaf Bossinger. 2003. "KLP-18, a Klp2 kinesin, is required for assembly of acentrosomal meiotic spindles in Caenorhabditis elegans." *Molecular Biology of the Cell* 14 (11):4458–69. doi:10.1091/mbc.E03-05-0283.

Sheng, J., K. Jiang, and X. Xue. 2019. "Knockdown of Kinase family 15 inhibits cancer cell proliferation in vitro and its Clinical relevance in Triple-Negative Breast Cancer." *Current Molecular Medicine*. doi:10.2174/1566524019666190308122108.

Shimamoto, Y., S. Forth, and T. M. Kapoor. 2015. "Measuring pushing and braking forces generated by ensembles of kinesin-5 crosslinking two microtubules." *Developmental Cell* 34 (6):669–81. doi:10.1016/j.devcel.2015.08.017.

Sleiman, Patrick M. A., Michael March, Kenny Nguyen, Lifeng Tian, Renata Pellegrino, Cuiping Hou, Walid Dridi, Mohamed Sager, Yousef H. Housawi, and Hakon Hakonarson. 2017. "Loss-of-function mutations in KIF15 underlying a Braddock–Carey genocopy." *Human Mutation* 38 (5):507–10. doi:10.1002/humu.23188.

Smertenko, A., S. L. Hewitt, C. N. Jacques, R. Kacprzyk, Y. Liu, M. J. Marcec, L. Moyo, A. Ogden, H. M. Oung, S. Schmidt, and E. A. Serrano-Romero. 2018. "Phragmoplast microtubule dynamics - a game of zones." *Journal of Cell Science* 131 (2). doi:10.1242/jcs.203331.

Song, X., T. Zhang, X. Wang, X. Liao, C. Han, C. Yang, K. Su, W. Cao, Y. Gong, Z. Chen, Q. Han, and J. Li. 2018. "Distinct diagnostic and prognostic values of kinesin family member genes expression in patients with breast cancer." *Medical Science Monitor* 24:9442–64. doi:10.12659/MSM.913401.

Sood, R., Y. Kamikubo, and P. Liu. 2017. "Role of RUNX1 in hematological malignancies." *Blood* 129 (15):2070–82. doi:10.1182/blood-2016-10-687830.

Stangeland, Biljana, Awais A. Mughal, Zanina Grieg, Cecilie Jonsgar Sandberg, Mrinal Joel, Ståle Nygård, Torstein Meling, Wayne Murrell, Einar O. Vik Mo, and Iver A. Langmoen. 2015. "Combined expressional analysis, bioinformatics and targeted proteomics identify new potential therapeutic targets in glioblastoma stem cells." *Oncotarget* 6:26192–215. doi:10.18632/oncotarget.4613.

Stock, M. F., J. Chu, and D. D. Hackney. 2003. "The kinesin family member BimC contains a second microtubule binding region attached to the N terminus of the motor domain." *Journal of Biological Chemistry* 278 (52):52315–22. doi:10.1074/jbc.M309419200.

Stockle, D., A. Herrmann, E. Lipka, T. Lauster, R. Gavidia, S. Zimmermann, and S. Muller. 2016. "Putative RopGAPs impact division plane selection and interact with kinesin-12 POK1." *Nature Plants* 2 (9). doi:Artn 1612010.1038/Nplants.2016.120.

Sturgill, E. G., and R. Ohi. 2013. "Kinesin-12 differentially affects spindle assembly depending on its microtubule substrate." *Current Biology* 23 (14):1280–90. doi:10.1016/j.cub.2013.05.043.

Sturgill, Emma G, Dibyendu Kumar Das, Yoshimasa Takizawa, Yongdae Shin, Scott E Collier, Melanie D Ohi, Wonmuk Hwang, Matthew J Lang, and Ryoma Ohi. 2014. "Kinesin-12 Kif15 targets kinetochore fibers through an intrinsic two-step mechanism." *Current Biology* 24 (19):2307–13. doi:10.1016/j.cub.2014.08.022.

Sturgill, Emma G., Stephen R. Norris, Yan Guo, and Ryoma Ohi. 2016. "Kinesin-5 inhibitor resistance is driven by kinesin-12." *Journal of Cell Biology* 213 (2):213–27. doi:10.1083/jcb.201507036.

Sueishi, M., M. Takagi, and Y. Yoneda. 2000. "The forkhead-associated domain of Ki-67 antigen interacts with the novel kinesin-like protein Hklp2." *Journal of Biological Chemistry* 275 (37):28888–92. doi:10.1074/jbc.M003879200.

Tanenbaum, M. E., L. Macurek, A. Janssen, E. F. Geers, M. Alvarez-Fernandez, and R. H. Medema. 2009. "Kif15 cooperates with eg5 to promote bipolar spindle assembly." *Current Biology* 19 (20):1703–11. doi:10.1016/j.cub.2009.08.027.

Toso, A., J. R. Winter, A. J. Garrod, A. C. Amaro, P. Meraldi, and A. D. McAinsh. 2009. "Kinetochore-generated pushing forces separate centrosomes during bipolar spindle assembly." *Journal of Cell Biology* 184 (3):365–72. doi:10.1083/jcb.200809055.

van Heesbeen, Roy G H. P., Marvin E Tanenbaum, and Rene H Medema. 2014. "Balanced activity of three mitotic motors is required for bipolar spindle assembly and chromosome segregation." *Cell Reports* 8 (4):948–56. doi:10.1016/j.celrep.2014.07.015.

van Oostende-Triplet, C., D. Guillet, T. Triplet, E. Pandzic, P. W. Wiseman, and A. Geitmann. 2017. "Vesicle dynamics during plant cell cytokinesis reveals distinct developmental phases." *Plant Physiology* 174 (3):1544–58. doi:10.1104/pp.17.00343.

Vanneste, D., M. Takagi, N. Imamoto, and I. Vernos. 2009. "The role of Hklp2 in the stabilization and maintenance of spindle bipolarity." *Current Biology* 19 (20):1712–7. doi:10.1016/j.cub.2009.09.019.

Varga, V., C. Leduc, V. Bormuth, S. Diez, and J. Howard. 2009. "Kinesin-8 motors act cooperatively to mediate length-dependent microtubule depolymerization." *Cell* 138 (6):1174–83. doi:10.1016/j.cell.2009.07.032.

Walker, K. L., S. Mueller, D. Moss, D. W. Ehrhardt, and L. G. Smith. 2007. "Arabidopsis TANGLED identifies the division plane throughout mitosis and cytokinesis." *Current Biology* 17 (21):1827–36. doi:10.1016/j.cub.2007.09.063.

Wang, Jie, Xingjun Guo, Chencheng Xie, and Jianxin Jiang. 2017. "KIF15 promotes pancreatic cancer proliferation via the MEK-ERK signalling pathway." *British Journal of Cancer* 117 (2):245–55. doi:10.1038/bjc.2017.165.

Wittmann, T., H. Boleti, C. Antony, E. Karsenti, and I. Vernos. 1998. "Localization of the kinesin-like protein Xklp2 to spindle poles requires a leucine zipper, a microtubule-associated protein, and dynein." *Journal of Cell Biology* 143 (3):673–85.

Wittmann, Torsten, Matthias Wilm, Eric Karsenti, and Isabelle Vernos. 2000. "Tpx2, a novel Xenopus map involved in spindle pole organization." *The Journal of Cell Biology* 149 (7):1405–18. doi:10.1083/jcb.149.7.1405.

Wolff, Ian D., Michael V. Tran, Timothy J. Mullen, Anne M. Villeneuve, and Sarah M. Wignall. 2016. "Assembly of Caenorhabditis elegans acentrosomal spindles occurs without evident microtubule-organizing centers and requires microtubule sorting by KLP-18/kinesin-12 and MESP-1." *Molecular Biology of the Cell* 27 (20):3122–31. doi:10.1091/mbc.e16-05-0291.

Xu, M., D. Liu, Z. Dong, X. Wang, Y. Liu, P. W. Baas, and M. Liu. 2014. "Kinesin-12 influences axonal growth during zebrafish neural development." *Cytoskeleton (Hoboken).* doi:10.1002/cm.21193.

Xu, X. F. M., Q. Zhao, T. Rodrigo-Peiris, J. Brkljacic, C. S. He, S. Muller, and I. Meier. 2008. "RanGAP1 is a continuous marker of the Arabidopsis cell division plane." *Proceedings of the National Academy of Sciences of the United States of America* 105 (47):18637–42. doi:10.1073/pnas.0806157105.

Yu, X., X. He, L. M. Heindl, X. Song, J. Fan, and R. Jia. 2019. "KIF15 plays a role in promoting the tumorigenicity of melanoma." *Experimental Eye Research.* doi:10.1016/j.exer.2019.02.014.

Zhao, H. D., Q. Y. Bo, Z. L. Wu, Q. G. Liu, Y. Li, N. Zhang, H. Guo, and B. K. Shi. 2019. "KIF15 promotes bladder cancer proliferation via the MEK-ERK signaling pathway." *Cancer Management and Research* 11:1857–68. doi:10.2147/Cmar.S191681.

Zou, J. X., Z. Duan, J. Wang, A. Sokolov, J. Xu, C. Z. Chen, J. J. Li, and H. W. Chen. 2014. "Kinesin family deregulation coordinated by bromodomain protein ANCCA and histone methyltransferase MLL for breast cancer cell growth, survival, and tamoxifen resistance." *Molecular Cancer Research* 12 (4):539–49. doi:10.1158/1541-7786. MCR-13-0459.

12 The Kinesin-16 Family

Hanan M. Alghamdi and Claire T. Friel

CONTENTS

The Kinesin-16 family is a group of kinesins found in organisms that build cilia or flagella.

12.1 EXAMPLE FAMILY MEMBERS

Mammalian: KIF12
Drosophila melanogaster: KLP54D
Trypanosoma brucei: KIF16A, KIF16B

12.2 STRUCTURAL INFORMATION

The only member of the Kinesin-16 family for which a body of data currently exists is the mammalian KIF12. The human KIF12 gene encodes a 651-amino acid protein, with an N-terminally positioned kinesin motor domain, followed by a predicted region of coiled coil with an internal hinge, and a C-terminal tail domain (Figure 12.1) (Katoh and Katoh 2005, Nakagawa et al. 1997). Human KIF12 shares 81% sequence identity with the KIF12 found in mouse, with both containing the characteristic kinesin motor domain including the four conserved nucleotide-binding motifs. There is no structure currently available for KIF12, but sequence similarity suggests that there will be high structural conservation with respect to motor domains from other kinesin families.

12.3 FUNCTIONAL PROPERTIES

There is little functional data currently available for the Kinesin-16 family. Phylogenetic analysis of the kinesin superfamily shows that Kinesin-16s are found only in organisms that build cilia or flagella (Wickstead, Gull, and Richards 2010),

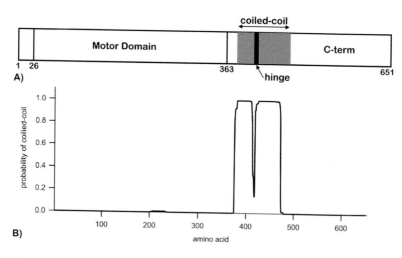

FIGURE 12.1 Domain layout for the mammalian Kinesin-16, KIF12. (A) Ribbon diagram showing the predicted domain layout of KIF12, numbered according to the amino acid sequence of human KIF12. (B) Probability of coiled-coil formation of the primary sequence of human KIF12, predicted using COILS v2.1.

suggesting a possible role in building the axoneme or in intra-flagellar transport (IFT). The domain layout of KIF12 is similar to that of translocating kinesins (Figure 12.1), which may suggest a cargo-carrying function; members of the Kinesin-2 family are known to function in this way in the building and maintenance of cilia (Scholey 2013). However, other cilium-associated kinesins appear to function by regulating microtubule dynamics (He et al. 2014, Kobayashi et al. 2011), and it is possible that KIF12 has a microtubule-regulating function rather than a translocating one.

12.4   PHYSIOLOGICAL ROLE

The mammalian KIF12 is the only Kinesin-16 for which the physiological role has been studied to date. KIF12 was first identified as a kinesin in a PCR (polymerase chain reaction) screen of a mouse cDNA library (Nakagawa et al. 1997). This study shows KIF12 to be predominantly expressed in the kidney. A more recent study indicates that KIF12 is also expressed in pancreatic islet cells (Yang et al. 2014). A quantitative transcriptomics analysis of all major human organs showed that KIF12 RNA is expressed at a high level in the kidney and at moderately high levels in the pancreas, gall bladder, thyroid gland and small intestine (Fagerberg et al. 2014).

Genetic analysis of mouse models of polycystic kidney disease (PKD) indicates that KIF12 is a modifier gene for PKD and has a major effect on the severity of the disease phenotype (Mrug et al. 2005, 2015). This possible role in PKD and evidence of localisation of KIF12 to the primary cilia in the IMCD3 cell line (Figure 12.2) supports the suggestion that Kinesin-16s are a family of cilium- and flagellum-associated kinesins (Mrug et al. 2015). KIF12 also appears to have a role in glucose metabolism and control of blood sugar levels. KIF12 knockout mice suffer from glucose intolerance due to reduced insulin secretion (Yang et al. 2014).

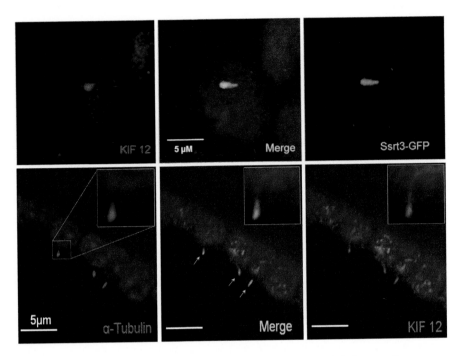

FIGURE 12.2 KIF12 localises to the primary cilium. Upper panel: Immunofluorescence images of KIF12 (red) co-localised with a green fluorescent protein (GFP)-tagged primary cilium marker, somatostatin receptor 3 (green), in a mIMCD cell line. Lower panel: Co-localisation of KIF12 (green) with α-tubulin (red), another primary cilium marker. Mrug 2015 PLOS One.

12.5 INVOLVEMENT IN DISEASE

The transcription factor hepatocyte nuclear factor-1β (HNF-1β) regulates expression of KIF12 in the kidney (Gong et al. 2009). Mutations of HNF-1β lead to a syndrome of inherited renal cysts and diabetes (Bellanne-Chantelot et al. 2004, Horikawa et al. 1997). In line with this, KIF12 has been identified as a candidate polycystic kidney disease (PKD) modifier gene, which has a major effect on the severity of the renal phenotype associated with this disease (Mrug et al. 2005). A five-amino acid deletion in the KIF12 protein results in a less severe phenotype than the full-length variant of KIF12 (Mrug et al. 2015).

A study of KIF12 knockout mice indicates a role in the regulation of glucose metabolism, possibly via an influence on insulin secretion, suggesting that dysfunction of KIF12 may be involved in the progression of certain types of diabetes (Yang et al. 2014). In support of this suggestion, the *KIF12* gene was one of six found to harbour potentially deleterious low- frequency mutations in a study of Type 2 diabetes in the Qatari population (O'Beirne et al. 2018).

Mutations in the *KIF12* gene have also been shown to be strongly correlated with paediatric cholestatic liver disease (Maddirevula et al. 2019, Unlusoy Aksu et al. 2019). This is a disease characterised by decreased bile flow due to impaired secretion by hepatocytes. The suggested association of KIF12 dysfunction with conditions such as cholestatic liver disease and reduced insulin secretion, together with its tissue distribution and localisation to the primary cilia, may suggest a general role for KIF12 in the function of secretory cells.

REFERENCES

Bellanne-Chantelot, C., D. Chauveau, J. F. Gautier, D. Dubois-Laforgue, S. Clauin, S. Beaufils, J. M. Wilhelm, C. Boitard, L. H. Noel, G. Velho, and J. Timsit. 2004. "Clinical spectrum associated with hepatocyte nuclear factor-1beta mutations." *Ann Intern Med* 140 (7):510–7. doi:10.7326/0003-4819-140-7-200404060-00009.

Fagerberg, L., B. M. Hallstrom, P. Oksvold, C. Kampf, D. Djureinovic, J. Odeberg, M. Habuka, S. Tahmasebpoor, A. Danielsson, K. Edlund, A. Asplund, E. Sjostedt, E. Lundberg, C. A. Szigyarto, M. Skogs, J. O. Takanen, H. Berling, H. Tegel, J. Mulder, P. Nilsson, J. M. Schwenk, C. Lindskog, F. Danielsson, A. Mardinoglu, A. Sivertsson, K. von Feilitzen, M. Forsberg, M. Zwahlen, I. Olsson, S. Navani, M. Huss, J. Nielsen, F. Ponten, and M. Uhlen. 2014. "Analysis of the human tissue-specific expression by genome-wide integration of transcriptomics and antibody-based proteomics." *Mol Cell Proteomics* 13 (2):397–406. doi:10.1074/mcp.M113.035600.

Gong, Y., Z. Ma, V. Patel, E. Fischer, T. Hiesberger, M. Pontoglio, and P. Igarashi. 2009. "HNF-1beta regulates transcription of the PKD modifier gene Kif12." *J Am Soc Nephrol* 20 (1):41–7. doi:10.1681/ASN.2008020238.

He, M., R. Subramanian, F. Bangs, T. Omelchenko, K. F. Liem, Jr., T. M. Kapoor, and K. V. Anderson. 2014. "The kinesin-4 protein Kif7 regulates mammalian Hedgehog signalling by organizing the cilium tip compartment." *Nat Cell Biol* 16 (7):663–72. doi:10.1038/ncb2988.

Horikawa, Y., N. Iwasaki, M. Hara, H. Furuta, Y. Hinokio, B. N. Cockburn, T. Lindner, K. Yamagata, M. Ogata, O. Tomonaga, H. Kuroki, T. Kasahara, Y. Iwamoto, and G. I. Bell. 1997. "Mutation in hepatocyte nuclear factor-1 beta gene (TCF2) associated with MODY." *Nat Genet* 17 (4):384–5. doi:10.1038/ng1297-384.

Katoh, M., and M. Katoh. 2005. "Characterization of KIF12 gene in silico." *Oncol Rep* 13 (2):367–70.

Kobayashi, T., W. Y. Tsang, J. Li, W. Lane, and B. D. Dynlacht. 2011. "Centriolar kinesin Kif24 interacts with CP110 to remodel microtubules and regulate ciliogenesis." *Cell* 145 (6):914–25. doi:10.1016/j.cell.2011.04.028.

Maddirevula, S., H. Alhebbi, A. Alqahtani, T. Algoufi, H. S. Alsaif, N. Ibrahim, F. Abdulwahab, M. Barr, H. Alzaidan, A. Almehaideb, O. AlSasi, A. Alhashem, H. A. Hussaini, S. Wali, and F. S. Alkuraya. 2019. "Identification of novel loci for pediatric cholestatic liver disease defined by KIF12, PPM1F, USP53, LSR, and WDR83OS pathogenic variants." *Genet Med* 21 (5):1164–72. doi:10.1038/s41436-018-0288-x.

Mrug, M., R. Li, X. Cui, T. R. Schoeb, G. A. Churchill, and L. M. Guay-Woodford. 2005. "Kinesin family member 12 is a candidate polycystic kidney disease modifier in the cpk mouse." *J Am Soc Nephrol* 16 (4):905–16. doi:10.1681/ASN.2004121083.

Mrug, M., J. Zhou, C. Yang, B. J. Aronow, X. Cui, T. R. Schoeb, G. P. Siegal, B. K. Yoder, and L. M. Guay-Woodford. 2015. "Genetic and informatic analyses implicate Kif12 as a candidate gene within the Mpkd2 locus that modulates renal cystic disease severity in the Cys1cpk mouse." *PLoS One* 10 (8):e0135678. doi:10.1371/journal.pone.0135678.

Nakagawa, T., Y. Tanaka, E. Matsuoka, S. Kondo, Y. Okada, Y. Noda, Y. Kanai, and N. Hirokawa. 1997. "Identification and classification of 16 new kinesin superfamily (KIF) proteins in mouse genome." *Proc Natl Acad Sci U S A* 94 (18):9654–9. doi:10.1073/pnas.94.18.9654.

O'Beirne, S. L., J. Salit, J. L. Rodriguez-Flores, M. R. Staudt, C. Abi Khalil, K. A. Fakhro, A. Robay, M. D. Ramstetter, J. A. Malek, M. Zirie, A. Jayyousi, R. Badii, A. Al-Nabet Al-Marri, A. Bener, M. Mahmoud, M. J. Chiuchiolo, A. Al-Shakaki, O. Chidiac, D. Stadler, J. G. Mezey, and R. G. Crystal. 2018. "Exome sequencing-based identification of novel type 2 diabetes risk allele loci in the Qatari population." *PLoS One* 13 (9):e0199837. doi:10.1371/journal.pone.0199837.

Scholey, J. M. 2013. "Kinesin-2: a family of heterotrimeric and homodimeric motors with diverse intracellular transport functions." *Annu Rev Cell Dev Biol* 29:443–69. doi:10.1146/annurev-cellbio-101512-122335.

Unlusoy Aksu, A., S. K. Das, C. Nelson-Williams, D. Jain, F. Ozbay Hosnut, G. Evirgen Sahin, R. P. Lifton, and S. Vilarinho. 2019. "Recessive mutations in KIF12 cause high gamma-glutamyltransferase cholestasis." *Hepatol Commun* 3 (4):471–7. doi:10.1002/hep4.1320.

Wickstead, B., K. Gull, and T. A. Richards. 2010. "Patterns of kinesin evolution reveal a complex ancestral eukaryote with a multifunctional cytoskeleton." *BMC Evol Biol* 10:110. doi:10.1186/1471-2148-10-110.

Yang, W., Y. Tanaka, M. Bundo, and N. Hirokawa. 2014. "Antioxidant signaling involving the microtubule motor KIF12 is an intracellular target of nutrition excess in beta cells." *Dev Cell* 31 (2):202–14. doi:10.1016/j.devcel.2014.08.028.

13 Other Kinesin Families

Claire T. Friel

CONTENTS

13.1 KINESIN-7

Of the Kinesin-7 family, only the kinetochore-associated kinesin CENP-E has been the focus of significant study. CENP-E is a processive, microtubule plus-end-directed translocating kinesin (Wood et al. 1997, Yardimci et al. 2008). The domain layout of CENP-E is equivalent to other plus-end-directed kinesins, with the motor domain located at the N-terminal end of the primary sequence, followed by a region of coiled-coil that facilitates dimerisation and a C-terminal tail domain. The ATP turnover cycle of CENP-E closely resembles that of a Kinesin-1 (Rosenfeld et al. 2009). However, CENP-E is slower and more processive and maintains microtubule attachment for long periods. The coiled-coil region of CENP-E is highly flexible, with a contour length almost three-fold longer than bovine brain Kinesin-1 (Kim et al. 2008). These characteristics adapt CENP-E for its role as a microtubule-to-kinetochore tether.

Depletion of CENP-E from a *Xenopus* egg extract disrupts chromosome alignment at the metaphase plate, and mitosis fails to arrest in response to spindle damage (Wood et al. 1997, Abrieu et al. 2000). CENP-E has been shown to transport chromosomes to microtubule plus ends and maintain association with dynamic microtubule ends at the metaphase plate (Shrestha and Draviam 2013, Gudimchuk et al. 2013), thereby playing a crucial role in chromosome congression and alignment. Due to this critical role in mitosis, variants of CENP-E result in developmental disorders such as microcephalic primordial dwarfism (Mirzaa et al. 2014).

13.2 KINESIN-9

The mammalian members of the Kinesin-9 family, KIF6 (KIF9B subgroup) and KIF9 (KIF9A subgroup), are N-terminal motors with a similar domain arrangement to the Kinesin-1 family. There is little functional information currently available for this family, but it consists only of sequences from species possessing cilia or flagella (Wickstead and Gull 2006).

Depletion of KIF9 in HeLa cells leads to slower progression through mitosis and an increased proportion of cells with misaligned chromosomes in metaphase (Andrieu et al. 2012). KIF9 localises to the mitotic spindle and knockdown of KIF9 results in an increased rate of spindle microtubule growth in cells. These data point to a role in regulating spindle dynamics. Data also suggest a role for KIF9 in macrophage activity: knockdown of KIF9 in macrophages reduces the number of podosomes to less than half the number present in control cells (Cornfine et al. 2011).

No functional data is currently available for KIF6. However, available physiological data suggests a role in ciliogenesis. Mice expressing truncated KIF6 display severe hydrocephalus (Konjikusic et al. 2018). Expression of KIF6 is shown to be specifically within the ependymal cells, which control cerebrospinal fluid flow in the ventricular system of the brain via ciliary beating. In *Xenopus*, epidermis KIF6 is shown to localise to the axoneme and basal body of multi-ciliated cells. Variation in KIF6 sequence is also suggested to influence not only an individual's risk of coronary heart disease, but also the response to statin treatment (Li et al. 2010, Ruiz-Iruela et al. 2018).

Trypanosomes also possess two Kinesin-9s, KIF9A and KIF9B, which both localise to the flagellum (Demonchy et al. 2009). Individual knockdown of these kinesins produces different motility defects. KIF9A is required for flagellar beating but is not involved in assembly, whereas KIF9B is suggested to form part of the machinery that regulates flagellar assembly.

13.3   KINESIN-17, -18, -19 AND -20

These four families have a more restricted species distribution than most other kinesin families (Wickstead, Gull, and Richards 2010). The Kinesin-17 family is found only in organisms that build cilia or flagella (Wickstead and Gull 2006). There is currently no data to illuminate the functional or physiological activity of Kinesin families 18–20.

REFERENCES

Abrieu, A., J. A. Kahana, K. W. Wood, and D. W. Cleveland. 2000. "CENP-E as an essential component of the mitotic checkpoint in vitro." *Cell* 102 (6):817–26. doi:10.1016/s0092-8674(00)00070-2.

Andrieu, G., M. Quaranta, C. Leprince, and A. Hatzoglou. 2012. "The GTPase Gem and its partner Kif9 are required for chromosome alignment, spindle length control, and mitotic progression." *FASEB J* 26 (12):5025–34. doi:10.1096/fj.12-209460.

Cornfine, S., M. Himmel, P. Kopp, K. El Azzouzi, C. Wiesner, M. Kruger, T. Rudel, and S. Linder. 2011. "The kinesin KIF9 and reggie/flotillin proteins regulate matrix degradation by macrophage podosomes." *Mol Biol Cell* 22 (2):202–15. doi:10.1091/mbc.E10-05-0394.

Demonchy, R., T. Blisnick, C. Deprez, G. Toutirais, C. Loussert, W. Marande, P. Grellier, P. Bastin, and L. Kohl. 2009. "Kinesin 9 family members perform separate functions in the trypanosome flagellum." *J Cell Biol* 187 (5):615–22. doi:10.1083/jcb.200903139.

Gudimchuk, N., B. Vitre, Y. Kim, A. Kiyatkin, D. W. Cleveland, F. I. Ataullakhanov, and E. L. Grishchuk. 2013. "Kinetochore kinesin CENP-E is a processive bi-directional tracker of dynamic microtubule tips." *Nat Cell Biol* 15 (9):1079–88. doi:10.1038/ncb2831.

Kim, Y., J. E. Heuser, C. M. Waterman, and D. W. Cleveland. 2008. "CENP-E combines a slow, processive motor and a flexible coiled coil to produce an essential motile kinetochore tether." *J Cell Biol* 181 (3):411–9. doi:10.1083/jcb.200802189.

Konjikusic, M. J., P. Yeetong, C. W. Boswell, C. Lee, E. C. Roberson, R. Ittiwut, K. Suphapeetiporn, B. Ciruna, C. A. Gurnett, J. B. Wallingford, V. Shotelersuk, and R. S. Gray. 2018. "Mutations in Kinesin family member 6 reveal specific role in ependymal cell ciliogenesis and human neurological development." *PLoS Genet* 14 (11):e1007817. doi:10.1371/journal.pgen.1007817.

Li, Y., O. A. Iakoubova, D. Shiffman, J. J. Devlin, J. S. Forrester, and H. R. Superko. 2010. "KIF6 polymorphism as a predictor of risk of coronary events and of clinical event reduction by statin therapy." *Am J Cardiol* 106 (7):994–8. doi:10.1016/j.amjcard.2010.05.033.

Mirzaa, G. M., B. Vitre, G. Carpenter, I. Abramowicz, J. G. Gleeson, A. R. Paciorkowski, D. W. Cleveland, W. B. Dobyns, and M. O'Driscoll. 2014. "Mutations in CENPE define a novel kinetochore-centromeric mechanism for microcephalic primordial dwarfism." *Hum Genet* 133 (8):1023–39. doi:10.1007/s00439-014-1443-3.

Rosenfeld, S. S., M. van Duffelen, W. M. Behnke-Parks, C. Beadle, J. Corrreia, and J. Xing. 2009. "The ATPase cycle of the mitotic motor CENP-E." *J Biol Chem* 284 (47):32858–68. doi:10.1074/jbc.M109.041210.

Ruiz-Iruela, C., A. Padro-Miquel, X. Pinto-Sala, N. Baena-Diez, A. Caixas-Pedragos, R. Guell-Miro, R. Navarro-Badal, X. Jusmet-Miguel, P. Calmarza, J. L. Puzo-Foncilla, P. Alia-Ramos, and B. Candas-Estebanez. 2018. "KIF6 gene as a pharmacogenetic marker for lipid-lowering effect in statin treatment." *PLoS One* 13 (10):e0205430. doi:10.1371/journal.pone.0205430.

Shrestha, R. L., and V. M. Draviam. 2013. "Lateral to end-on conversion of chromosome-microtubule attachment requires kinesins CENP-E and MCAK." *Curr Biol* 23 (16):1514–26. doi:10.1016/j.cub.2013.06.040.

Wickstead, B., and K. Gull. 2006. "A 'holistic' kinesin phylogeny reveals new kinesin families and predicts protein functions." *Mol Biol Cell* 17 (4):1734–43. doi:10.1091/mbc.e05-11-1090.

Wickstead, B., K. Gull, and T. A. Richards. 2010. "Patterns of kinesin evolution reveal a complex ancestral eukaryote with a multifunctional cytoskeleton." *BMC Evol Biol* 10:110. doi:10.1186/1471-2148-10-110.

Wood, K. W., R. Sakowicz, L. S. Goldstein, and D. W. Cleveland. 1997. "CENP-E is a plus end-directed kinetochore motor required for metaphase chromosome alignment." *Cell* 91 (3):357–66. doi:10.1016/s0092-8674(00)80419-5.

Yardimci, H., M. van Duffelen, Y. Mao, S. S. Rosenfeld, and P. R. Selvin. 2008. "The mitotic kinesin CENP-E is a processive transport motor." *Proc Natl Acad Sci U S A* 105 (16):6016–21. doi:10.1073/pnas.0711314105.

14 Summary and Future Perspectives

Claire T. Friel

It is 35 years since the discovery of a protein with the ability to move microtubules and which was named kinesin. In the intervening time, kinesin has been shown to be not a unique molecule, but the founding member of a superfamily that is ubiquitous across eukaryotes. The common feature that defines a protein as a member of the kinesin superfamily is the motor domain, which binds both ATP and microtubules, and confers the nucleotide-dependent interaction with microtubules which is characteristic of all kinesins. Classification of the superfamily is based on phylogenetic analysis of motor domain sequences and the sequence of its motor domain is sufficient to place a kinesin into a particular family. In this book, we collate the structural and functional information available for each kinesin family, to allow the reader to quickly obtain an overview of the type of behaviour typical of each. Despite the high degree of structural conservation of the kinesin motor domain, family-specific motifs and regions of secondary structure exist that adapt members of particular families to different functions. This, combined with the diversity of non-motor regions found both within and across families, allows members of the kinesin superfamily to perform a wide range of physiological functions.

In addition to cargo transport functions (Kinesin-1, -2 and -3), many kinesins are involved in cell division and play crucial roles in the formation and functioning of the microtubule spindle that powers chromosome segregation, and in processes that drive cytoplasmic separation into daughter cells (Kinesin-4, -5, -6, -7, -8, -13, -14 and -15). Abnormal expression of kinesins involved in mitotic cell division is frequently found in various types of cancer and many kinesins are considered targets for anti-cancer therapies.

Kinesins are also crucial to the formation and function of two other major microtubular structures, the cilium and the flagellum, and certain kinesin families are found only in species that possess these structures (Kinesin-2, -9, -16 and -17). Members of the Kinesin-2 family are critical to intraflagellar transport, in which cargos are transported within cilia or flagella. Other kinesins regulate the structure and length of cilia and flagella by their ability to control microtubule dynamics (Kinesin-4, -8, -9 and -13). Mutations in kinesins involved in cilium growth and maintenance often result in diseases termed ciliopathies or in developmental disorders.

Neural cells require close control of microtubule organisation for the formation and functioning of neural processes, such as axons and dendrites. Many kinesins have roles in the development and function of neural cells (Kinesin-1, -2, -3, -4, -6, -13 and -15) and mutations in these kinesins can result in various neuropathies.

There remains tremendous scope for the study of kinesins and the variety of cellular processes they support. For example, the role of kinesins in plant cell structure and function is only just beginning to be uncovered. Also, very little is known about the role of kinesins in immune cells. Certain kinesins have been shown to play a role in macrophage function (Kinesin-3, -4 and -9), but little is known about the molecular mechanisms involved. There are several kinesin families for which little or no structural, functional or physiological information is currently available (Kinesin-7, -9, -16, -17, 18, -19 and -20). The set of information on individual kinesin families collated in this book will hopefully provide an overview of the existing knowledge across the superfamily and provide a basis from which to spot not only the gaps in this knowledge, but also the avenues deserving of further study.

Index

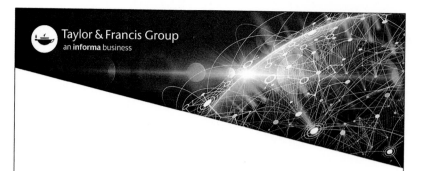